RAND ARROYO CENTER

Modern Political Warfare

Current Practices and Possible Responses

Linda Robinson, Todd C. Helmus, Raphael S. Cohen, Alireza Nader, Andrew Radin, Madeline Magnuson, Katya Migacheva

Prepared for the United States Army

For more information on this publication, visit www.rand.org/t/RR1772

Library of Congress Control Number: 2018938057
ISBN: 978-0-8330-9707-1

Published by the RAND Corporation, Santa Monica, Calif.

Cover: iStock/Getty Images Plus

Preface

This study, *Modern Political Warfare: Current Practices and Possible Responses*, examines political warfare as practiced today by both state and nonstate actors, including Russia, Iran, and the Islamic State. The study recommends revisions in U.S. military and nonmilitary approaches and capabilities to better address threats short of conventional warfare. The findings and recommendations should be of interest to the U.S. military, the U.S. State Department, and those in the executive and legislative branches charged with national security policy responsibilities. Allied and wider public audiences may also find this study of interest.

The study was produced as the deliverable for an Army project titled "Theory and History of Political Warfare (POLWAR)." The project sponsor requested that RAND Arroyo Center examine whether a modern variant of political warfare is needed to address its potential use as a tool to confront ideological threats, as well as its potential use by nonstate or state-sponsored entities. The sponsor also requested examination of implications for Army special operations forces and interagency and multinational partners. The interviews and research for this report were conducted in 2016 and 2017.

This research was sponsored by the U.S. Army Special Operations Command G-9 and conducted within the RAND Arroyo Center's Strategy, Doctrine, and Resources Program. RAND Arroyo Center, part of the RAND Corporation, is a federally funded research and development center sponsored by the United States Army.

The Project Unique Identification Code (PUIC) for the project that produced this document is RAN156926.

Contents

CHAPTER SIX

CHAPTER SEVEN

Figures and Tables

Figures

Tables

Summary

Study Objective and Research Approach

The United States today faces a number of actors who employ a wide range of political, informational, military, and economic measures to influence, coerce, intimidate, or undermine U.S. interests or those of friends and allies. The objective of this study is to provide a clearer view of these adversarial measures short of conventional warfare and to derive implications and recommendations for the U.S. government and U.S. military. Toward this end, RAND conducted a yearlong examination of the historical and current practices that fall into this realm of conflict short of conventional war. The starting point for this examination is the term "political warfare," as defined in 1948 at the outset of the Cold War by U.S. diplomat George Kennan:

> Political warfare is the logical application of Clausewitz's doctrine in time of peace. In broadest definition, political warfare is the employment of all the means at a nation's command, short of war, to achieve its national objectives. Such operations are both overt and covert. They range from such overt actions as political alliances, economic measures (as ERP—the Marshall Plan), and "white" propaganda to such covert operations as clandestine support of "friendly" foreign elements, "black" psychological warfare and even encouragement of underground resistance in hostile states.

Political warfare is but one term among many that describes the arena of conflict short of conventional warfare. Chinese analysts have employed the term "unrestricted warfare," Russian officials have used "soft power" and "new generation warfare," and a variety of terms are in use by U.S. officials, including "gray zone conflicts," "hybrid warfare," "asymmetric warfare," and "irregular warfare." The latter term has been officially defined in U.S. military doctrine and Department of Defense (DoD) directives, but one impetus for a new nomenclature is to place emphasis on the nonmilitary and nonlethal elements of this form of warfare. Those elements may be readily combined with conventional warfare, but the focus of this examination is on the less obvious, more ambiguous forms of conflict that may catch policymakers unaware if they are insufficiently attuned to these methods and their abilities to sow conflict, weaken, destabilize, disrupt, and, in some cases, create more dramatic consequences, as seen in Russia's rapid annexation of the Crimea without resorting to all-out warfare. This examination of political warfare does not presuppose that this term is necessarily the most apt appellation for current nonconventional contests of power, but it employs the term as a matter of historical record and convenience to bound the study to the nonmilitary and nonlethal military methods used. By the same token, measures short of war may be usefully employed to deter conflict or prevent it from escalating or worsening.

The research approach for this study includes a review of historical accounts and documents that reflect the U.S. experience with political warfare, an investigation of primary sources to assess the current practices of political warfare employed in three case studies, and extensive semi-structured interviews with practitioners and experts. We supplement this research with official documents and other source material to evaluate U.S. and Allied responses, capability gaps, and possible remedies for meeting these challenges short of war.

Organization of Study

The study opens with a historical chapter that examines the U.S. experience and use of political warfare methods, which were developed and employed most fully during the Cold War. To consider and characterize modern-day strategies and practices that might qualify as political warfare, as conducted by state and nonstate actors alike, the research team's subject matter experts consider four cases studies. Two cases examine current strategies and practices by state actors—Russia and Iran—and one case examines practices of a nonstate actor—the Islamic State in Syria and the Levant (ISIL). While China is also considered a prominent example of many of these practices, the scope of the study limited the number of case studies to two state actors and one nonstate actor. Selected references are included for China, and the study team recommends that an in-depth study of its current practices be conducted to supplement this work.

These detailed case studies provide the basis for a synthesis chapter that identifies the principal characteristics or attributes of modern political warfare. This synthesis, in turn, provides the basis for considering (a) information operations currently in use to respond to this form of warfare, (b) gaps in current U.S. governmental practices or capabilities to respond to this form of warfare, and (c) recommendations for more effective employment of all of the national elements of power in a coordinated, "whole-of-government" manner.

U.S. Historical Use of Political Warfare

While U.S. political warfare came into vogue during the Cold War, such practices were employed early in the country's history, even before its founding. During the Cold War, a wide variety of political, cultural, informational, and economic measures were employed through institutions such as the United States Information Agency (USIA), the U.S. Agency for International Development (USAID), the Department of State (DoS), and the Central Intelligence Agency (CIA). Among the measures that helped repudiate Communist ideology were material efforts to support authors, artists, and magazines as well as Voice of America broadcasts. Economic measures included long-term secret assistance to countries, political parties, and unions or other movements. Covert and clandestine measures included assassinations and other "dirty tricks," which were restricted following the Church and Pike Committee investigations. These hearings led President Gerald

Ford to issue Executive Order 11905 in 1976, explicitly banning political assassination. Subsequent executive orders restated this prohibition, effectively bounding the types of activities that U.S. entities were permitted to legally conduct. After the Cold War, many institutions such as USIA were dismantled, its public diplomacy officers were folded into the State Department, and funding and personnel in both USAID and DoS were dramatically reduced. In the 1990s, some effort was made to revive certain capabilities. In addition to retaining overseas radio and television broadcasting stations, for example, the U.S. government developed new democracy promotion and economic programs through the National Endowment for Democracy and the Millennium Challenge Corporation.

Russia Case Study

This case study examines contemporary Russian thought and tactics in the realm of political warfare, as practiced primarily against the Baltic states. The Estonian "Bronze Soldier" incident of 2007 is analyzed in depth, using primary sources. The case study introduces a schema for understanding Russia's employment of nonstate entities in a more or less concerted approach to political warfare. It also finds that the effectiveness of the tactics vary greatly depending on the environment in which they are applied.

Specific findings include the following:

- Russia views its activities as defensive in reaction to the United States. Russia considers U.S. democracy promotion, support for civil society, and its open media environment as highly threatening political warfare tools wielded by the state.
- In Estonia, Russia demonstrated an opportunistic rather than deliberate approach to what it calls "new generation warfare." In other words, Russia used political warfare by capitalizing on crises. Russian shaping operations (e.g., propaganda directed at Russian speakers) prepared the ground, followed by more overt tools of political warfare applied at opportune moments.

- Innovations in Russian "new generation warfare" are economic leverage, social proxies, and media penetration.
- Russia has invested heavily in propaganda, but the principal effect of its extensive media operations may be to obfuscate through disinformation rather than to gain converts through persuasion.

Iran Case Study

Iran employs a wide suite of measures short of conventional war to advance its regional aims. Much of the official guidance concerning these practices can be documented in senior officials' speeches. While a great deal of extant literature focuses on the Iranian Republican Guard Corps-Qods Force (IRGC-QF), this case study details the extensive political, cultural, informational, and economic programs that extend Iran's influence and partnerships around the world. The study conducts a detailed examination of Iran's development and use of militias in Iraq and Syria, which employ Arab forces to reach Arab audiences and serve as conduits for religious and political influence as well as military effects.

Specific findings include the following:

- Iran "soft power" strategy is heavily based on cultural, political, and religious influence as primary and powerful levers. Iran employs a targeted, differentiated approach to Shi'a, pan-Arab, and pan-Islamic audiences.
- Iran utilizes a worldwide network of cultural, informational, and influence organizations, backed by material support. The IRGC-QF is only one element of that support.
- Iran offers political and economic support to foreign political parties and leaders to install and influence governments.
- Iranian religious tactics include funding of junior clerics and mass pilgrimages.
- Iran employs a two-stage use of militias: Many Iranian proxy forces mature into political actors, requiring a continual regeneration of proxy forces.

- The use of Arab proxies in Syria (including Iraqi militias and Lebanese Hezbollah paramilitaries) is as important to Iran as direct intervention.

Islamic State Case Study

ISIL has gained notoriety for its brutal practices and ability to conduct maneuver warfare with equipment seized from Iraqi and Syrian forces. However, it has also distinguished itself by an astute and innovative use of measures other than military might, including social media (to recruit fighters and promote its vision) and an array of political and economic tactics that extend its control and intimidate populations. This case study surveys ISIL's foundational documents and its suite of nonlethal tactics. It also conducts a lexical analysis of its social media content on Twitter to discern key differences in its Arabic- and English-language approaches to information operations.

Specific findings include the following:

- ISIL announced a "Caliphate" primarily to unify fighters under a single banner and to enhance recruitment.
- ISIL systematically indoctrinates new recruits to strip them of their old identities and to prevent them from straying.
- ISIL media adaptations have limited the damage posed by its suspension from Twitter; ISIL has moved away from a broadcast model to a dispersed and resilient form of communication that relies on peer-to-peer sharing and redundancy across platforms.
- ISIL targets different audiences with different messages; ISIL employs violence and emotive language liberally in Arabic-media productions to mobilize rank-and-file members, while English-language channels are more restrained and "international" in focus.

Characteristics of Modern Political Warfare

The research team's synthesis of the political warfare practices identified in the case studies above produced a list of key attributes that broadly describe how this form of warfare is conducted today. This analysis also suggests the limits of this form of warfare; that is, it is not effective in all environments, particularly where the adversary lacks credibility or leverage with key audiences and where a society is politically, economically, and socially resilient.

In summary form, these attributes are the following:

- Nonstate actors can conduct political warfare with unprecedented reach.
- Political warfare employs all the elements of national power.
- Political warfare relies heavily on unattributed forces and means.
- The information arena is an increasingly important battleground, where perceptions of success can be determinative.
- Information warfare works in various ways by amplifying, obfuscating, and, at times, persuading. Compelling evidence supplied in a timely manner is the best antidote to disinformation.
- Detecting early-stage political warfare requires a heavy investment of intelligence resources.
- Political warfare can generate unintended consequences.
- Economic leverage is increasingly the preferred tool of the strong.
- Political warfare often exploits shared ethnic or religious bonds or other internal seams.
- Political warfare extends, rather than replaces, traditional conflict and can achieve effects at lower cost.

Furthermore, in considering the utility of the term "political warfare" to describe a U.S. policy approach, the research team interviewed over 40 practitioners and experts and found that the term was not considered ideal for a wide variety of reasons. Chief among them were the objections that many government entities do not, as a matter of rule, conduct warfare, and that their efforts would be less effective if branded as such. Another objection held that informational campaigns in particular would carry less weight with intended audiences if branded as warfare. Yet another detractor noted that the global information space today is characterized primarily by peer-to-peer platforms and conversation, debate, and dialogue rather than one-way exhortations and messaging resonant of wartime communications or propaganda. While political warfare has a historical pedigree and descriptive value, the overwhelming judgment of those consulted led the research team to recommend against employing the term as a description of what the United States does or should do. Rather, a concluding chapter refers instead to "effective statecraft" and, in particular, to an "integration of measures short of war" as necessary aspects of the U.S. arsenal. The final chapter refers to efforts to deter, disrupt, and defeat this type of intervention as "countering political warfare." An additional category of action is steps taken to reduce vulnerabilities and increase resilience.

Influence Communications

This chapter considers the information activities of the State Department and the U.S. military, particularly psychological operations or what are now called Military Information Support Operations (MISO). The importance of the information space—and the ability to operate effectively within that space—stands out as a major feature of conflicts short of war. This area must therefore be considered and managed at the highest levels of government, given its ability to profoundly affect all other lines of effort. Moreover, the revolution in communications and information technology has transformed the information space, necessitating new models and new capabilities to compete effectively in this arena.

The research findings identified several gaps in information capabilities and practices in the U.S. government. The gaps include the following:

- Strategic-level communications are high-profile and bureaucratically risky, characteristics that militate against speed and initiative.

- The new Global Engagement Center (GEC), established by presidential executive order and located at the DoS, focuses on third-party validators or influencers from the bottom up, whereas MISO within DoD remains more focused on government top-down communications.
- MISO is challenged by significant manpower shortages and limited new media training.
- U.S. Central Command (CENTCOM), with the aid of contractors, is at the forefront of DoD social media communications; other combatant commands (COCOMS) are lagging.
- Interagency coordination and National Security Council (NSC) guidance pertaining to message themes remains lacking.

Several implications for U.S. policymakers, and for the U.S. special operations community in particular, flow from these findings:

- Leadership must empower communicators to quicken response times and take risks.
- Unattributed communications may have counterproductive effects that should be anticipated and mitigated.
- The use of third parties is a critical requirement for both the GEC and MISO.
- MISO requires both increased manpower and new media training.
- GEC and DoS must improve relations.
- Interagency coordination remains a fundamental challenge.

Effective Statecraft and Integration of Measures Short of War

An integrated response to threats short of conventional war include (1) the need for strategy, (2) the need for a whole-of-government approach to statecraft led by an appropriately enabled State Department, (3) the formulation and coordination of responses with and through other sovereign governments, allies, and partners, and (4) recommendations for the improvement of military contributions to such an integrated approach.

It is beyond the scope of the study to define current U.S. strategic interests in detail, but the general requirement for a cost-effective approach to national defense suggests that early and effective nonmilitary responses, and nonlethal uses of its military element of national power, may provide the U.S. government with valuable tools to deter adversaries, prevent conflicts from escalating, or mitigate their effects. In some cases, these approaches may effectively reduce or remove the incipient threats. A positive formulation for U.S. strategy can also increase the deterrent effect imposed on aggressive or revisionist actors and attract greater support among those who stand to benefit. Thus, rather than a purely reactive stance, the U.S. strategy can be formulated as promoting and defending a rules-based international order (including measures to reform and enhance those rules for the benefit of all) while defending U.S. interests and values.

The State Department is the designated lead for the conduct of U.S. foreign policy and the representation of its interests abroad. As such, it is the logical entity to lead a whole-of-government response in a realm that is primarily political and diplomatic, and to coordinate other agencies if given such policy guidance from the President. The in-depth knowledge of career Foreign Service officers regarding the history and current dynamics of specific countries and regions is unmatched in the U.S. government. In contrast, the NSC staff, which has grown enormously in the past two decades, still lacks the depth and experience of the State Department, since the NSC is staffed by appointees and rotating personnel from the departments and agencies. Despite the deep country and regional expertise resident at the State Department, an in-depth series of interviews with current and retired senior diplomats and other experts, supplemented by recent studies by the American Foreign Service Association and the American Academy of Diplomacy, highlight significant gaps in organizational and operational capabilities and practices that should be remedied to enable the State Department to effectively plan, coordinate, and execute interagency responses in a continuous manner, if so directed by the President.

The study also considers the ability of the U.S. government to work effectively with other countries in addressing these challenges. U.S. plans and activities must necessarily be coordinated with the governments of those countries where the aggression, subversion, coercion, or destabilization is occurring, along with other partners or allies who are willing and able to contribute their resources and efforts in a common effort. The examples of Estonia and NATO are explored as mini-cases to illustrate requirements and general principles for effective response. The example of Great Britain is also presented to derive lessons regarding further adaptation of policy, military doctrine, military organization, and personnel solutions for responding to asymmetric and hybrid threats. In particular, the UK army has initiated novel ways to develop nonlethal and nonmilitary capabilities that draw together military and civilian expertise.

Finally, the study concludes with eight recommendations for improving the practices and capabilities of the U.S. military and of U.S. special operations forces (SOF) in particular to combat nonconventional warfare through expanded deterrence, enhanced resilience, and preparations for national resistance, among other means. In some cases, the U.S. Army Special Operations Command and its higher headquarters, U.S. Special Operations Command, may be able to make adaptations deemed beneficial; in other cases, such decisions would rest with the Department of Defense.

Recommendations

Recommendation 1: To improve whole-of-government synergy, U.S. military commands, to include deployed headquarters, should as a matter of course include civilian departmental representatives in order to understand, coordinate with, and support U.S. State Department and other civilian program execution.

Recommendation 2: The Defense Department, and special operations forces in particular, should incentivize and improve selection and training for military advisers serving at State Department headquarters, U.S. embassies, and other diplomatic posts in order to increase their effectiveness.

Recommendation 3: The Defense Department, and special operations forces, should offer military planners to the State Department as the latter builds its own cadre of planners and integrates regional and functional bureau plans, to enable State to play a lead role in responses to political warfare.

Recommendation 4: Military commanders should develop and maintain collaborative relationships with their civilian counterparts through regular visits and frequent communications to develop common understanding of and approaches to political-military conflict.

Recommendation 5: DoD should routinely seek to incorporate State Department knowledge and the current insights of the U.S. country team into military plans in order to develop effective responses to political-military threats.

Recommendation 6: The special operations community should make it a high priority to improve and implement fully resourced, innovative, and collaborative information operations.

Recommendation 7: Military commanders and the State Department should identify critical information requirements for gray zone threats, and the intelligence community should increase collection and analysis capabilities that are dedicated to detecting incipient subversion, coercion, and other emerging threats short of conventional warfare.

Recommendation 8: DoD and State should support deployment of special operations forces as an early and persistent presence to provide assessments and develop timely and viable options for countering measures short of conventional war. Consideration should be given to likelihood of exposure of sensitive, clandestine, or covert activities.

Acknowledgments

The RAND project team wishes to acknowledge the support of the sponsor, LTG Kenneth Tovo, his command group, and the G-9 led by Mr. Robert Warburg. In particular, Len Gaddis and Dr. Robert Toguchi of the G-9 provided consistent support and consultation during this yearlong project. The project leader, Linda Robinson, also benefited from the USASOC Silent Quest 16 series of events, which included in-depth discussions of political warfare.

The study team received voluminous input from many U.S., European, and other allied officials as well as subject matter experts who have studied political warfare and its analogues since the Cold War. We wish to thank those who hosted us in various European venues on multiple trips, provided us tours of critical areas, and participated in the February 2016 "Measures Short of War" conference cosponsored with our RAND Europe colleagues in Cambridge, England. The non-attribution basis of our interviews prevents us from thanking individuals by name, but we wish to express particular gratitude to senior serving and retired members of the diplomatic corps who regard this subject as very much germane to practitioners of the entire spectrum of statecraft.

Finally, the RAND team wishes to thank its peer reviewers, Dr. Kathleen Hicks, senior vice president; Henry A. Kissinger, chair, and director of the International Security Program at the Center for Strategic and International Studies; and RAND senior researcher Burgess Laird, for extensive, probing critiques and recommendations that greatly improved this work. We also thank our many colleagues at RAND who provided reviews, suggestions, and assistance in the course of this study, to include Dr. Sally Sleeper, director of Arroyo Center's Strategy, Doctrine and Resources program, her predecessor Dr. Terrence Kelly, Communications Analyst John Godges, Terri Perkins, Darlette Gayle, Rhonda Normandin, Donna White, Francisco Walter, Julienne Amsden, and the highly professional RAND Publications Department.

Abbreviations

AAH	Asa-ib Ahl al-Haq
ABA	Ahl al-Bayt Assembly
ABM	Ansar Beit al-Maqdis
AC	alternative current
ACO	Allied Command Operations
AFRICOM	Africa Command
AP	Associated Press
BARB	British Broadcasters Audience Research Board
BBC	British Broadcasting Corporation
CBT	Center for a Better Tomorrow
CENTCOM	Central Command
CIA	Central Intelligence Agency
CIS	Commonwealth of Independent States
CMSE	Civilian Military Support Element
COCOM	Combatant Command
COMINTERN	Communist International

CSCC	Center for Strategic Counterterrorism Communications
CSO	Community Service Officer
CSTO	Collective Security and Treaty Organization
DC	direct current
DDoS	Distributed Denial of Service
DIME	Diplomatic, Informational, Military, and Economic
DNS	Domain Name System
DoD	Department of Defense
DoS	Department of State
EDL	Estonian Defense League
EEU	Eurasian Economic Union
ERP	European Recovery Program
EU	European Union
EUCOM	European Command
FAPSI	Federal Agency of Government Communication and Information (Russia)
FARC	Revolutionary Armed Forces of Colombia
FNLA	Front for the Liberation of Angola
FSB	Federal Security Service (Russia)
FSO	Federal Protection Service (Russia)
GEC	Global Engagement Center
GPD/LD	General Political Department Liaison Department (China)

GRU	Main Intelligence Directorate of the General Staff (Russia)
ICA	Islamic Cyber Army
ICC	Iranian Cultural Center
ICRO	Islamic Culture and Relations Organization
ICSR	International Centre for the Study of Radicalisation and Political Violence
IDC	Institute for Democracy and Cooperation
IIA	Information and Influence Activities
IRGC	Islamic Revolutionary Guards Corps
IRGC-QF	Islamic Revolutionary Guards Corps-Qods Force
IRIB	Islamic Republic of Iran Broadcasting
ISC	Islamic State Caliphate
ISCI	Islamic Supreme Council in Iraq
ISI	Islamic State of Iraq
ISIL	Islamic State of Iraq and the Levant
JAM	Jaish al-Mahdi
JIATF	Joint Interagency Task Force
JTAC	Joint Terminal Air Controller
KAPO	Kaitsepolitsei (Estonia)
KGB	*Komitet gosudarstvennoy bezopasnosti*
KH	Kata'ib Hezbollah
KIA	Katabi al-Imam Ali
LAFA	Liwa Abu Fadl al-Abbas
LH	Lebanese Hezbollah

LICHR	Legal Information Center for Human Rights
LTTE	Liberation Tigers of Tamil Eelam
MA	military assistance
MANPAD	Man-Portable Air Defense System
MBN	Mir Bez Nazisma (World Without Nazism)
MISO	Military Information Support Operations
MIST	Military Information Support Team
MOOTW	Military Operations Other Than War
NATO	North Atlantic Treaty Organization
NCL	Non-Communist Left
NDF	National Defense Force
NGO	Nongovernmental Organization
NSC	National Security Council
OIF	Operation Iraqi Freedom
OPC	Office of Policy Coordination
OSS	Office of Strategic Services
OWI	Office of War Information
PD	public diplomacy
PMU	popular mobilization unit
POLWAR	political warfare
PSYOPS	psychological operations
PUK	Patriotic Union of Kurdistan
RFE	Radio Free Europe
RFE/RL	Radio Free Europe/Radio Liberty
RL	Radio Liberty

RT	Russia Today
SACEUR	Supreme Allied Commander Europe
SCIRI	Supreme Council for the Islamic Revolution in Iraq
SOCOM	Special Operations Command
SOF	Special Operations Forces
SOFLE	Special Operations Forces Liaison Element
SOST	Special Operations Support Team
SOUTHCOM	Southern Command
SSA	Security Sector Assistance
SVR	Foreign Intelligence Service (Russia)
UAE	United Arab Emirates
UIA	United Iraqi Alliance
UK	United Kingdom
UNITA	National Union for the Total Independence of Angola
USAID	United States Agency for International Development
USASOC	U.S. Army Special Operations Command
USIA	United States Information Agency
UW	unconventional warfare
VBIED	vehicle-borne improvised explosive device
VDV	*Vozdushno-desantnye voiska*
VOA	Voice of America

CHAPTER ONE

Introduction

> Political warfare is the logical application of Clausewitz's doctrine in time of peace. In broadest definition, political warfare is the employment of all the means at a nation's command, short of war, to achieve its national objectives. Such operations are both overt and covert. They range from such overt actions as political alliances, economic measures (as ERP—the Marshall Plan), and "white" propaganda to such covert operations as clandestine support of "friendly" foreign elements, "black" psychological warfare and even encouragement of underground resistance in hostile states.
>
> —*George F. Kennan, U.S. diplomat and historian, May 4, 1948*

Purpose of Study

Gray zone conflicts, hybrid warfare, irregular warfare, and other similar terms have been applied to a dizzying array of old and new asymmetric tactics in contests for power that are roiling many regions. This study examines this phenomenon through the lens of a somewhat antiquated term—"political warfare"—as defined above by former U.S. diplomat George Kennan to focus squarely on measures short of conventional combat, how they are employed today, and what responses might be effective for the United States and its allies to apply.

Two rationales motivate this approach to exploring contemporary conflict and the means that democratic states may use to address it. The first rationale is that the United States may have devoted too little attention of late to understanding the ways in which actors are employing a host of nonkinetic tactics and how they form part of the spectrum of conflict, giving these actors a relative advantage in their expansive approach to competition for power and influence. The second rationale follows from this, which is to explore whether the United States might leverage all its national (and other) sources of power and influence more effectively to protect and advance its interests.

What Is "Political Warfare"?

The term "political warfare" is often attributed to famed American diplomat George Kennan, who wrote about the phenomenon at the early stages of the Cold War. In a May 4, 1948, policy memorandum, Kennan—then head of the State Department's policy planning staff—drafted the statement cited at the forefront of this chapter.

Kennan, whose career straddled the diplomatic and intelligence worlds, envisioned a far-reaching battle for influence—waged below the threshold of traditional state-on-state conflict—using all the levers of national power (Corke, 2006, p. 105). Kennan viewed the struggle between the United States and the Soviet Union as not simply a matter of state competition, but a battle of ideologies between liberal democracy and Communism, and political warfare became one of the primary means to conduct this new form of war (Kennan to the Secretary of State, 1946).[1]

Political warfare never fit neatly in American strategic dialogue, however. Many American military officers originally dismissed the term because it sounded too British (Corke, 2006, p. 109). In fact, Kennan's original policy memorandum specifically notes political warfare's Anglophone roots: "The creation, success, and survival of the British Empire has been due in part to the British understanding and application of the principles of political warfare" ("On the Inauguration of Organized Political Warfare," 1948, para. 2).

---

[1] For Kennan's depiction of the ideological war against the Soviet Union, see his "long telegram."

Commentators have also raised at least three substantive objections to the term. First, some have objected to the word "political" (Hoffman, 2014). As Clausewitz famously remarked, "war is the continuation of politics by other means," which implies that *all* wars—not just some subgroup—are political wars (Clausewitz, 1984, p. 87). Second, others have flinched at applying the term "warfare" to behavior done without a formal declaration of hostilities, against both friends and adversaries, where physical violence is often not employed (Hoffman, 2014). Finally, Kennan's definition can include practically every form of state interaction—short of full-on state-on-state conventional war—and since the latter is a historically rare phenomenon, the term is not particularly useful. Some historians suggest that Kennan chose the term "political warfare" only so that the State Department—not the military or the intelligence community—was responsible for this arena (Corke, 2006, p. 109). In his original 1948 memorandum, Kennan boldly stakes his claim: "Both [overt and covert political warfare], from their basic nature, should be directed and coordinated by the Department of State" ("On the Inauguration of Organized Political Warfare," 1948, para. 4).

Thanks to these definitional problems, many strategists have discarded "political warfare" entirely in favor of other terms. As Professor Carnes Lord remarked, "The problem is indicated by the general tendency to use the terms *psychological warfare* and *political warfare* interchangeably to designate the overall phenomenon, not to mention a variety of similar terms—ideological warfare, the war of ideas, political communication and more" (Lord, 1989, p. 16). French lieutenant colonel and counterinsurgency strategist David Galula explicitly equated the two terms, "revolutionary war is a political war," and argued that in both, "politics becomes an active instrument of operation" (Galula, 1964, pp. 6–7). Similarly, French colonel Roger Trinquier used the term "subversive warfare" in almost the same manner as Kennan used "political warfare": to describe "an interlocking system of actions political, economic, psychological, military—that aims at the overthrow of the established authority in a country and its replacement by another regime" (Trinquier, 1964, p. 6). A Central Intelligence Agency (CIA) doctrinal manual from the 1980s proclaimed that "guerilla warfare is essentially a political war" ("Tayacán," 1985, p. 33). More recently, senior defense officials and experts including National Defense University's Frank Hoffman have suggested that "hybrid warfare" may be a better term than political warfare (Hoffman, 2014). However, as commonly used, hybrid warfare connotes a combination of nonkinetic or irregular military tactics with conventional military means, rather than a combination of military and civilian tools to achieve effects. The preponderance of tools in political warfare are nonkinetic in nature, and their combination with conventional military means constitutes the "hybrid" aspect. While hybrid warfare is certainly a phenomenon of current war, this study is focused on the political warfare aspect rather than the admixture.

Other strategists have tried to give "political warfare" more precision than Kennan's original definition. Historian Paul Smith, for example, tries to tie the term to psychological warfare:

> Political war is the use of political means to compel an opponent to do one's will, *political* being understood to describe purposeful intercourse between peoples and governments affecting national survival. Political war may be combined with violence, economic pressure, subversion, and diplomacy, but its chief aspect is the use of words, images, and ideas, commonly known, according to context, as propaganda and psychological warfare. (Smith, 1989, p. 3)

Angelo Codevilla takes Smith's argument that political warfare is largely about psychological warfare and gives it a more succinct definition, similar to that of political action. "Political warfare is the marshaling of human support, or opposition, in order to achieve victory in a war or in unbloody conflicts as serious as war" (Codevilla, 1989, p. 77).

And yet, while there is widespread agreement that psychological warfare is an aspect of political warfare, most view political warfare as a broader, more encompassing term (Lucas and Mistry, 2009, p. 40). Carnes Lord and Frank Barnett concluded in their edited volume on the subject, "Political warfare is a term that is less well established in usage and doctrine, but one that seems useful for describing a spectrum of overt and covert activities designed to support national political-military objectives" (Barnett and Lord, 1989, p. xi). In Lord and

Barnett's definition, political warfare represents a basket of tactics—not just propaganda—and while certainly not as simple and probably less intellectually satisfying, their definition is, perhaps, more in keeping with Kennan's original conception, which combined psychological warfare with other tools that are used to achieve political ends (Lucas and Mistry, 2009, p. 48). The question, then, is what tactics belong under political warfare's umbrella.

Lord provides a partial answer: "Political warfare is a general category of activities that includes political action, coercive diplomacy, and covert political warfare" (Lord, 1989, p. 18). He then defines political action as support for foreign political parties, coercive diplomacy as diplomacy backed up by the threat of military force, and covert political warfare as the aid of guerilla movements, black propaganda, and influence operations (Lord, 1989, p. 18).

More recently, the United States Army Special Operations Command's March 2015 white paper asserts that "Political Warfare encompasses a spectrum of activities associated with diplomatic and economic engagement, Security Sector Assistance (SSA), novel forms of Unconventional Warfare (UW), and Information and Influence Activities (IIA)" (United States Army Special Operations Command, 2015, p. 2). Like Lord's definition, the white paper's definition emphasizes influence operations and unconventional warfare. Unlike Lord's definition (and perhaps unsurprisingly, given its audiences), the Army white paper focuses more on military tools—less on diplomacy and political tools.

Based on the manifold reactions to the term, we can begin to sketch out what political warfare is and is not. On the one hand, political warfare includes all the tools of national power: diplomatic, informational, military, and economic. Depending on the case, some of these tools may have greater or lesser relevance, but all of them are possible tools of political warfare.

On the other hand, just as importantly, the reactions hint at what political warfare is not—namely, neither regular diplomatic and economic interactions nor full-on-conventional war between states. Political warfare exists in a gray zone between these two ends of the spectrum, although drawing the line between the two can also prove difficult. For example, while Lord seemingly includes overt political assistance and messaging as part of political warfare, others counter that such public diplomacy—overt attempts by sovereign states to sway foreign public opinion—lacks a coercive element to it and so should not be considered political warfare (Smith, 1989, p. 7). Conversely, historian Anne Karalekas defines political warfare precisely as "direct intervention in the electoral process of sovereign governments, rather than attempts to influence public opinion through the activities of the media" (Corke, 2006, p. 109).

Given the contrasting yet overlapping definitions of the term, it is best to view "political warfare" as a series of concentric circles with certain tactics falling more or less neatly under the overall rubric (see Figure 1.1).

Underlying all of the definitions is the presumption that "political warfare" is a deliberate tool of statecraft. Trade between nations may ebb and flow; information may cross borders; diplomatic ties may go up and down for a variety of reasons apart from deliberate policy choices. These shifts may, in turn, have political second-order effects. And yet, political warfare involves a deliberate policy choice to undermine a rival or achieve other explicitly political objectives by means other than routine diplomacy or all-out war. With this definition in mind, we turn, in the subsequent four chapters, to how the United States, Russia, Iran, and ISIL have practiced this form of warfare. We examine how these actors practice political warfare, with each chapter examining the specific contributions from the diplomatic/political, informational, military, and economic realms shown in Figure 1.1.

Research Approach and Organization of Study

The research team set out to answer three questions:

- What is political warfare?
- How is it (or an appropriate analogous term) waged today?
- How might the U.S. government and its allies and partners most effectively respond to or engage in this type of conflict to achieve their ends and protect their interests?

Figure 1.1
Defining the Contours of Political Warfare

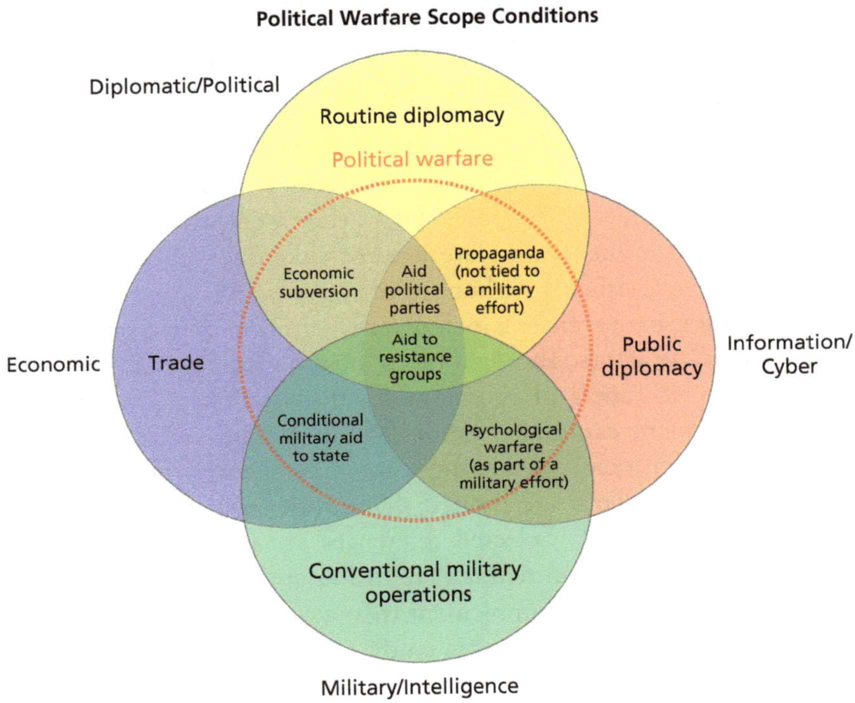

NOTE: Political warfare consists of the intentional use of one or more of the implements of power (diplomatic, information, military, and economic) to affect the political composition or decision-making within a state. Political warfare is often—but not necessarily—carried out covertly, but must be carried out outside the context of traditional war.
Caveat: All activities here are meant to be illustrative, rather than an exhaustive list of possible actions.
RAND *RR1772-1.1*

The research approach included a deeply researched historical chapter (Chapter Two) examining how the United States practiced political warfare in the past, particularly during the Cold War, followed by four case studies examining the conduct of what might be deemed modern political warfare. Using primary sources, the research team identified the doctrinal or theoretical basis for political warfare and its dominant measures or means used by Russia, Iran, and the Islamic State in Iraq and the Levant (ISIL), as well as corporate practices today that might be deemed an analogous form of corporate warfare. This research also explored the ability of nonstate actors to conduct this form of warfare. The case studies on Russia, Iran, and ISIL all considered the tools available to those actors under the commonly employed Diplomatic, Informational, Military, and Economic (DIME) construct, with additional attention paid to such areas as cyber and intelligence services insofar as they were used to carry out activities to gain influence and leverage or weaken, destabilize, subvert, or overthrow governments without resorting to war.

The case selection criteria included the selection of two state actors and two nonstate actors. The decision to focus on Russia and Iran, both prominent practitioners of the activities discussed here, meant that another prominent candidate, China, could not be included. While China is also considered a prominent example of many of these practices, the scope of the study limited the number of cases to two state and one nonstate cases. Selected references are included for China, and the study team recommends that an in-depth study of its current practices be conducted to supplement this work. The decision to focus on nonstate actors as part of this study corrects a tendency to view political warfare as solely the province of state actors. In the current era, empowered nonstate actors have the ability to employ many of the tools and tactics once reserved to states. Thus, the study team, with the support of the sponsor, chose to devote significant time to investigating current nonstate practices. These judgments do not diminish the importance of China as a prominent state actor deserving of in-depth study. The one-year parameter of this study dictated that only four in-depth cases could be conducted as part of the overall project.

The three case studies and additional research provided the baseline for a synthesis chapter (Chapter Seven) that identifies prominent attributes of modern political warfare and refines the understanding of how state and nonstate actors function in this spectrum of conflict and competition. The term "political warfare" is also evaluated for its utility and appropriateness in the current context. The final two chapters consider how the United States might grapple with this type of low-grade but persistent and pernicious form of conflict. Chapter Eight is devoted entirely to the realm of influence, information, and communications, given the prominence of these activities and the enormous changes in tactics due to the ongoing information technology revolution.

Chapter Nine evaluates the strategic, organizational, and other resources required by the U.S. government today to respond effectively to political warfare and similar challenges. The U.S. State Department, as the U.S. entity charged with representing and advancing U.S. foreign policy interests abroad, receives particular attention. The roles of the host countries, U.S. allies, and U.S. military, particularly the special operations forces (SOF), are also considered in evaluating the requisite coordination in planning and execution of nonmilitary activities. The particular focus on SOF, specifically their desired roles and needed capabilities, not only addresses the sponsor's requirement for such an evaluation but takes into account the wide range of SOF capabilities that can be brought to bear in nonwar environments. The perennial, and neuralgic, question of who does what and how everyone can best operate together is answered through the team's analysis of extensive interviews with a wide range of governmental actors. Chapter Ten offers final, overarching conclusions.

CHAPTER TWO

History of U.S. Political Warfare

On October 26, 1775, more than six months before the signing of the Declaration of Independence, the U.S. Continental Congress wrote to the people of Quebec, the French-speaking province of Canada. The pamphlet—later translated into French, printed in a pamphlet, and distributed among the population—aimed to convince them to join the colonies in revolt against the British crown and "complete the bright and strong chain of union" stretching from Nova Scotia to Georgia (Cohen, 2011, p. 134). The pamphlet praised Québécois military prowess for resisting British domination when Quebec was still a French colony. It highlighted the benefits and rights guaranteed under the new American union: freedom of religion, habeas corpus, property rights, and the like. Finally, it threatened, "You have been conquered into liberty, if you act as you ought. This work is not of man. You are a small people, compared to those who with arms invite you to fellowship. A moment's reflection should convince you which will be most for your interest and happiness, to have all the rest of North-America your unalterable friends, or your inveterate enemies" (Cohen, 2011, p. 134). Ultimately, the pamphlet failed to incite the French Canadians to join the colonies' cause. Still, it illustrated an early propensity to employ nonmilitary means to pursue objectives against a more powerful adversary.

The account above indicates that the American tradition of political warfare began long before the phrase itself was ever coined. In fact, political warfare is as old as recorded war itself. In this chapter, we first explore both the development of the term "political warfare" and its traditions specifically in American strategic thinking and practice. The remainder of this chapter is divided into three sections. The first section briefly describes how the United States conducted political warfare long before the term was ever coined, from the very beginning of the republic. The second section examines American political warfare's coming of age during the Cold War and depicts how the United States wielded diplomatic, information, military, and economic tactics as part of political warfare. The third section describes more recent American uses of political warfare.

American Political Warfare's Early History

Although it has not always been labeled as such, political warfare has been a part of American strategy since the very beginning. As mentioned previously, the United States encouraged French Canadians to revolt against the British Crown even before the United States itself became independent. This attempt at political warfare did not end with the American Revolution. In fact, the Articles of Confederation—the early document governing the United States—included an overt diplomatic overture to Canada to join the union. Article XI stated, "Canada acceding to this confederation, and joining in the measures of the united states, shall be admitted into, and entitled to all the advantages of this union: but no other colony shall be admitted into the same, unless such admission be agreed to by nine states" ("Transcript of Articles of Confederation [1777]," n.d.).

Economic subversion also proved a key element of eighteenth- and nineteenth-century American military history. Depending on how loose a definition of political warfare one adopts, the American boycott of British goods and the Boston Tea Party—where American revolutionaries famously threw British tea into Boston harbor—could count as political warfare during the American Revolution. Similarly, American patriot seizures of loyalist property (and vice versa) during the war itself would also fall into this basket of economic political warfare.

One of the most important tools of economic political warfare through the first half of the nineteenth century, however, revolved around the slave trade. Abolishing slavery threatened to upend the social order in Southern states and proved a powerful political tool. Even the British tried to leverage this weakness. As early as November 14, 1775, John Murray, Earl of Dunmore and royal governor of Virginia, offered freedom to any slave belonging to an American revolutionary in exchange for fighting with the loyalist forces. Some 800 former slaves joined Murray, while an estimated 10,000 are believed to have tried (PBS, "Proclamation of Earl Dunmore, 1775," n.d.). And while it was likely driven more from moral conviction, Abraham Lincoln's Emancipation Proclamation almost ninety years later, arguably, served a similar purpose of undermining the political order in the South.

Later in the nineteenth century, as the United States developed into a regional economic power, it began to use multinational corporations as means of political power. The classic example here is the United Fruit Company in Latin America. One historian recounts that "the United Fruit Company virtually ran Central America. In 1870, 'the octopus,' as Central Americans called it, was a fine cover for U.S. intelligence work. Its buyers, overseers, agents, and managers were all over the area by 1910, interested without apology in everything from local soil conditions to the president's latest mistress" (Omang, 1985, p. 5).

The twentieth century saw new advances in political warfare techniques employed as part of conventional conflicts, driven by both world wars. On the information front, President Woodrow Wilson created the Committee on Public Information in August 1917 under George Creel—with support of the Army, Navy and State Department—as a means to build American support for the war and counter enemy subversion at home (Barnett, 1989, p. 202). The United States took to political action abroad as well. Given the multiethnic composition of the Ottoman and Austro-Hungarian empires, backing local ethnic nationalism under these regimes became an attractive method to undermine both. In this light, some argue that American and British support for Czechoslovak nationalism during World War I constituted a form of political warfare to undermine the Austro-Hungarian Empire (Smith, 1989, p. 117). Backing nationalist movements could also have second-order international political benefits. Indeed, one of the reasons given for why Great Britain issued the Balfour declaration in 1917—backing a Jewish homeland in what was then Palestine—was to encourage Zionist movements in the United States and Russia to pressure their governments to back Zionist war aims (Vereté, 1970, pp. 55–57).

World War II presented an even greater arena for American political warfare. Arguably, the American entry into the war was itself brought on by a failed attempt at political warfare. In response to Japanese military aggression in China and Southeast Asia, the United States had imposed a full embargo against Japan, restricting its access to militarily vital oil, iron, and steel supplies; had frozen Japanese assets in American banks; and had ramped up military aid to Chinese forces fighting against Japanese troops. Economically cornered, Japan then attacked the American naval base in Pearl Harbor on December 7, 1941 (Office of the Historian, n.d.).

During the war, the United States embraced political warfare in all its forms. On June 13, 1942, the United States stood up the Office of Strategic Services (OSS) under the leadership of Major General William Donovan. Viewed as the forerunner to the Central Intelligence Agency, the OSS eventually numbered 14,000 employees with a budget of $45 million. It performed intelligence gathering as well as covert action tasks, including political warfare ("What Was OSS?," 2008). Together with the other Allied forces, the OSS sent "Jedburgh" teams—typically, three-man teams dropped behind enemy lines—to train the French, Belgian, and Dutch resistance ("Jedburghs," n.d.). The OSS also sent "Detachment 101" to Burma to organize the Kachin tribes and to Thailand to organize the "Free Thai" movement to fight against the Japanese. The OSS also sent "operational groups"—commando units—to France, Italy, Greece, Yugoslavia, Burma, Malaya, and China to fight with local partisans ("The Office of Strategic Services" 2012; Paddock, 2002, pp. 26–29). The OSS did not limit itself to kinetic action. It also conducted "morale operations" to promote resistance to the Axis and undermine enemy morale (Godson, 2008, p. 28).

The conduct of political warfare during World War II was not limited to the OSS. The Army conducted one of the largest and most successful guerilla operations of the war—training some 37,000 Filipino guerillas on the island of Mindanao (Paddock, 2002, p. 30–31). The Army also stood up the Psychological Warfare Branch in 1942—

renamed the Psychological Warfare Division in February 1944—to conduct "the dissemination of propaganda designed to undermine the enemy's will to resist, demoralize his force and sustain the morale of our supporters" (Paddock, 1989, p. 46). By the end of the war, the Psychological Warfare Division grew to impressive proportions, publishing eight newspapers with circulations of 1 million in Germany and another 2 million to displaced persons (Paddock, 2002, p. 41).

The United States stood up the Office of War Information (OWI) to aid its propaganda effort. According to the *Report on the Activities of the Office of War Information in the European Theater of Operations During Calendar Year 1944*, OWI "does not try" to "persuade people to like the United States; it tried to help people understand the United States, on the assumption that the more the truth about America is known, the more the nature of American civilization is understood, the better for all concerned" (quoted in Speier, 1948, p. 7). In truth, OWI tried to influence audiences both in the United States and abroad.

Importantly, during the World Wars, the United States viewed political warfare as a component of what were largely conventional wars, not as an independent type of warfare. Even on this model, however, political warfare is used as means to achieve ends without resorting to conventional methods. Nonetheless, at the end of World War II, the United States dismantled many of its instruments of political warfare, along with the general demobilization of its conventional forces. The OWI and the OSS were disbanded in 1945 ("What Was OSS?," 2008). The Psychological Warfare Division was renamed the Information Control Division and focused primarily on denazification (Paddock, 2002, p. 41). Still, by the end of World War II, the United States had gained the knowledge and capacity to conduct large-scale political warfare. It was not long after the war's end that the United States would once again employ these capabilities.

The Cold War and Political Warfare's Coming of Age

During the Cold War, political warfare came into its own as a tool of American statecraft, not just as the political, economic, and intelligence aspects of conventional war but as a separate domain in its own right. Given the prospect that conventional war could escalate into nuclear conflict, the premium increased on furthering aims without use of conventional force and often in a clandestine or covert manner. The set of DIME measures could be—and were—undertaken in any of three manners: overt, clandestine, or covert, depending on the choices of policymakers. The following sections address the overt information activities of the U.S. Information Agency and its component broadcasting entities, as well as overt political and economic measures.

Faced with the prospect of confronting the Soviet Union, the United States identified in political warfare a way to combat the ideological threat of Communism while staying below the threshold of full-scale war. Consequently, on June 18, 1948, President Harry Truman signed NSC 10/2 and created the CIA's Office of Special Projects, later the Office of Policy Coordination (OPC), to oversee the "conduct of secret political, psychological and economic warfare together with preventative direct action (paramilitary activities)—within the policy direction of the Departments of State and Defense" (Lynn, 2013, p. 20). From 1949 until 1952 (when it was folded back into the directorate of operations), OPC grew tenfold in personnel, from 302 to 3,142, and almost twenty-fold in terms of budget, from $4.7 million in 1949 to $82 million in 1952 (Saunders, 2013, p. 34).

Political warfare later made it into NSC 68 as one of the key tools—along with conventional forces—that the United States should employ in fighting the Cold War (Lucas and Mistry, 2009, p. 63). NSC 68 called for the "intensification of affirmative and timely measures and operations by covert means in the fields of economic warfare and political and psychological warfare with a view to fomenting and supporting unrest and revolt in selected strategic satellite countries" ("NSC 68," n.d. [1950]). NSC 68 helped lead to the quadrupling of the $34 million 1950 psychological warfare budget over the next two years (Saunders, 2013, p. 82). For the first half of the Cold War in particular, American policymakers were enthralled with the idea of covert action: in the mid-1950s, covert action consumed almost 50 percent of the CIA's budget (Woodward, 1987, p. 51).

The Vietnam War, Watergate scandals, and Church and Pike Committee investigations into the intelligence community's practices dulled some of the appeal of covert action (Turner, 1985, p. 83). During the 1976 investigations into the intelligence community's abuses of authority, Senator Frank Church argued that Congress never intended to give the CIA authority to conduct covert political warfare abroad (*Final Report of the Select Committee*, 1976, p. 563). By the mid-1970s, covert action was estimated at only 4 percent of the CIA's budget, and by 1990, it was estimated at a mere 1 percent (Woodward, 1987, p. 51; Daugherty, 2006, p. 34). While the Church Committee's report marked an inflection point, it did not bring an end to American political warfare. The use of covert action figured at the forefront of President Ronald Reagan's war against the "evil empire" (Reagan, 1983). Political warfare also continued in overt forms. Even in the waning days of the Cold War, the National Defense University published two book-length monographs on the subject of "political warfare," highlighting the continued academic interest in this subject (Smith, 1989; Barnett and Lord, 1989).

Throughout the Cold War, America's exercise of political warfare continued to span all four sides of the concentric circles constituting the "contours of political warfare," as illustrated in Figure 1.1 and as outlined below: diplomatic/political, information, military, and economic.

Diplomatic/Political

America's use of political aid proved remarkably successful during the Cold War. Two cases stand out in particular: one to aid anti-Communist groups in Western Europe soon after World War II, the other to aid the Solidarity movement in Poland decades later.

In the wake of the devastation of World War II, Communist parties in Western Europe flourished. In response, the United States began a covert program to aid anti-Communist labor unions, political parties, student groups, and media, particularly in France and Italy (Turner, 1985, p. 76). Starting with NSC 4/A, President Harry Truman approved up to $20 million to undermine "Soviet and Soviet-inspired activities which constitute a threat to world peace and security"

(Daugherty, 2006, p. 118). The logic was that by creating a viable left wing, but not Communist alternatives, one could sap political support from Communism. By 1967, the CIA spent approximately $10 million on such aid to foreign "friendly groups," often funneled through U.S. labor unions and student groups who acted as "cut outs" for the aid (Turner, 1985, p. 76).

This aid produced mixed results. During the 1948 elections, the CIA helped fund the Italian Christian Democrats' campaign. The mission was successful insofar as the Christian Democrats were elected, but once in office, they broke with the United States on policy. Consequently, some American policymakers wanted to kick these parties off the dole. U.S. ambassador to Italy Clare Booth Luce argued that "after five years and half billion dollars in overt aid and large amounts of covert aid, the leaders of the 'democratic parties' are telling us today, just as they told us in 1948, the only alternative to the Communist victory at the polls is for us to keep them and their parties on a permanent handout basis" (Mistry, 2006, p. 319). Despite Luce's frustration, the Italian Communist party failed to win elections from 1948 through the termination of the covert aid program in 1967 (Daugherty, 2006, p. 83, 160). More importantly, Western Europe remained in the Western orbit.

A relatively proactive example of political warfare came toward the end of the Cold War with U.S. support to the Polish labor union and political party, Solidarity. Run covertly by the CIA, with the help of the National Endowment for Democracy, the American Federation of Labor-Congress of Industrial Organization, the Polish American Congress, and others, the United States funneled between $10 and $50 million to the Solidarity movement during the 1980s (Fischer, 2012, pp. 429, 431). The money underwrote propaganda for the movement and supplied Solidarity with printers, copiers, and even an underground television broadcasting capability (Fischer, 2012, p. 432).

The operation proved successful. Despite the fact that the Soviet-backed regime declared martial law, some 300 book titles and 800 periodicals, mostly associated with Solidarity, began to circulate in Poland during the first year alone (Fischer, 2012, p. 437). The program only grew from there. To give some sense of the scale, Polish customs officials on November 29, 1986, found hidden in a truck, on a ferry from Sweden, 23 modern Western duplicator machines, 16 fax machines, 49 photocopiers, 420 cans of ink, 5,000 pamphlets and other books, IBM and Tandy computers, radio scanners, and radio transmitters (Fischer, 2012, p. 449).[1] The effort eventually contributed to Solidarity's electoral success as well. In the 1989 Polish elections, Solidarity won 160 of 161 seats in the lower chamber (the Sejm) and 92 of 100 seats in the upper chamber (the Senate) that were allowed to be contested (Fischer, 2012, p. 450).[2] Solidarity's cofounder Lech Wałęsa later went on to become Poland's second president, and Solidarity's victory in Poland played a significant role in hastening the collapse of the Soviet empire (Fischer, 2012, p. 455).

Informational

Information, propaganda, and psychological warfare efforts also figured prominently during the Cold War. The United States sponsored a series of studies to look at psychological warfare. In October 1950, the State Department asked the Massachusetts Institute of Technology to solve the problem of "getting truth behind the Iron Curtain" (Needell, 1993, p. 400).[3] Called Project Troy, the idea was to build on the propaganda efforts of World War II and the Office of War Information (Needell, 1993, p. 401). The study looked primarily at radio broadcasts and balloons, but also at mail, journals, student exchanges, travelers, and movies to get U.S. information to Central and Eastern Europe (Needell, 1993, p. 409).

The United States, ultimately, tried many of these means of propaganda. Between October 1951 and November 1956, some 350,000 balloons carried an estimated 300 million "friendship" leaflets, posters,

---

1 Importantly, while the CIA did not acknowledge whether this shipment specifically came from them per se, it did acknowledge some of its shipments in 1986 were seized by Polish officials (ibid.).

2 The Communist regime reserved 65 percent of the seats in the Sejm for Communist candidates.

3 In this sense, it was similar to an earlier President's Advisory Commission on Universal Training report of May 29, 1947, which concluded that "truth is our greatest ally" in curbing the spread of Communism (quoted in Speier, 1948, p. 10).

and books to Poland, Czechoslovakia, and Hungary. According to one 1957 survey of Hungarian defectors, 15 percent of them relied on these leaflets for news (Lynn, 2013, p. 42). Refugees in the West formed another tie to countries behind the Iron Curtain, forming a more informal network to funnel information back to their home countries (Lynn, 2013, p. 60).

One of the more creative means of Cold War propaganda came in the form of culture. Especially in the early period, the CIA funded cultural activities to the tune of tens of millions of dollars a year, "in effect acting as America's Ministry of Culture" (Saunders, 2013, p. 108). The CIA bankrolled the Congress for Cultural Freedom, an organization that funded authors and artists around the world to promote "the fundamental ideals governing cultural (and political) action in the Western world and the repudiation of all totalitarian challenges" (Center for the Study of Intelligence, 2008). The organization owed its name to a congress convened in West Berlin's Titania Palace on June 26, 1950 (by happenstance the day after North Korea invaded the South), with some 4,000 cultural representatives present, including prominent Americans such as the playwright Tennessee Williams and the historian Arthur Schlesinger, Jr. (Center for the Study of Intelligence, 2008). After the meeting, the congress later lived on as a grant-making organization. The CIA also funneled money through 170 different foundations (Saunders, 2013, p. 113). One study of 164 charitable foundations found that, of the 700 grants over $10,000 made between 1963 and 1966, 108 were partially or wholly funded by the CIA (Saunders, 2013, p. 112).

The cultural endeavors funded through these efforts ran the gamut. The CIA promoted displays of abstract expressionist art to counter Soviet realism. As one account described, abstract expressionism "spoke to a specifically anti-Communist ideology, the ideology of freedom, of free enterprise. Nonfigurative and politically silent, it was the very antithesis to socialist realism" (Saunders, 2013, p. 213). Perhaps most significantly, the CIA also backed a series of magazines—*Der Monat* for German speakers (Smith, 1989, p. 202), *Perspectives*, and, most notably, *Encounter* (Saunders, 2013, pp. 117, 147). As *Perspectives* founder James Laughlin said, the magazines aimed "not so much to *defeat* the leftist intellectuals in dialectical combat as to *lure* them away from their positions by aesthetic and rational persuasion" (Saunders, 2013, p. 117 [emphasis in original]). *Encounter* became particularly successful. It featured some of the best-known authors of the day, including Julian Huxley, W. H. Auden, Thornton Wilder, Robert Penn Warren, and others (Saunders, 2013, p. 179). *Encounter* also attracted a wide readership for its day: the first 10,000 copies of its first issue sold out quickly, and it eventually rose to become the CIA's "flagship" journal of this kind (Saunders, 2013, p. 155, 185).

The United States military also developed a psychological operations capability during the Cold War. At the end of World War II, much of the Army's psychological warfare capability disbanded, only to be reconstituted during the Korean War to perform both tactical and strategic missions (Paddock, 1989, p. 47). Over the next several decades, the fortunes of the military's psychological operations community ebbed and flowed with strategic conditions, downsizing in the early 1960s, then ramping up during Vietnam, only to decline again in subsequent years (Paddock, 1989, pp. 48–49). By the mid-1980s, in response to a presidential directive, the military increased the attention paid to enhancing its ability to conduct psychological operations both for large-scale wars and in more limited interventions (Paddock, 1989, p. 50).

Perhaps the centerpiece of the U.S. information effort, however, was more overt in nature. In 1953, the United States founded the United States Information Agency (USIA). Its mission was to

- explain and advocate U.S. policies in terms that are credible and meaningful in foreign cultures;
- provide information about the official policies of the United States, and about the people, values, and institutions that influence those policies;
- bring the benefits of international engagement to American citizens and institutions by helping them build strong long-term relationships with their counterparts overseas; and
- advise the President and U.S. government policymakers on the ways in which foreign attitudes will have a direct bearing on the effectiveness of U.S. policies. (Nakamura and Weed, 2009, p. 10)

In its heyday, USIA sponsored publications in 30 languages, sent hundreds of speakers abroad, and administered a variety of cultural and professional exchanges (Nakamura and Weed, 2009, p. 10). USIA administered Voice of America (VOA), which grew out of the Office of War Information and served as "the official overseas broadcast arm of the United States" (Nakamura and Weed, 2009, p. 9; "VOA History," n.d.). In 1999, USIA ran 190 USIS posts, staffed by 520 USIA Foreign Service Officers and augmented by 2,521 locally hired Foreign Service Nationals, spread across 142 countries (Nakamura and Weed, 2009, p. 12).

Other parts of America's information effort were Radio Free Europe and Radio Liberty, which came directly under USIA's purview later in its existence. On August 26, 1948, CIA, OPC, State and Defense Department officials met to discuss a radio broadcasting service for Eastern Europe (Lynn, 2013, p. 22). This led to the creation of two organizations, Radio Free Europe (RFE), broadcasting to Central and Eastern Europe; and Radio Liberty (RL), targeting the Soviet Union itself. While often lumped together, RFE and RL were separate operations. RFE was a network of organizations, each with its own director, research unit, and programming staff, composed mostly of émigrés. RL, by contrast, was smaller, a single station, and focused mainly on broadcasting in Russian with smaller minority-language desks (Smith, 1989, p. 199). Both RFE and RL became large-scale efforts. Of the $356.9 million that the United States gave to Free Europe, Inc.—the umbrella organization that supported European émigré activities—from May 1949 to June 30, 1971, about 70 percent went to fund Radio Free Europe (Lynn, 2013, p. 27).[4]

Whether RFE/RL should count as propaganda is debatable. Some argue that, unlike their Soviet counterpart Radio Moscow, RFE/RL adhered to journalistic standards of "objectivity and balance" (Lord, 1989, p. 26). Still, RFE/RL played a key role in broadcasting "truth" behind the Iron Curtain. According to some accounts, RFE/RL helped inspire an uprising of German workers in 1953 and both Polish unrest

[4] According to some experts, this also was about the time when RFE/RL reached their peak as tools of political warfare (see Smith, 1989, p. 198).

and a Hungarian revolt in 1956 (Rowen, 1989, p. 184). Unfortunately, the broadcasts may have also created the false expectation of U.S. support for these uprisings. One International Research Associates study of Hungarian refugees in 1957 found that 96 percent of them had expected Western aid and 77 percent of them had expected military aid in 1956 (Lynn, 2013, p. 46). Still, RFE/RL continued to serve as an alternate source of news, particularly for events downplayed or ignored by official Soviet news sources. For example, RFE/RL became an important source of information for Ukraine on the Chernobyl nuclear disaster (Rowen, 1989, p. 184). By the end of the Cold War, an estimated 60 percent of the Soviet population listened to Western radio, many of this segment under the age of 30 (Alexiev, 1989, p. 201).

Military

Perhaps the most dramatic, if controversial, examples of American political warfare during the Cold War were more kinetic actions—where the United States assisted in the overthrow of left-wing regimes suspected of collaborating with the Soviet Union or aiding insurgent forces. Importantly, while the military aspects often overshadowed the nonlethal dimensions, these operations were often combined with substantial nonlethal components, including the use of some of the information and political tools described above. And while many of these operations were tactically quite successful, they tended to leave checkered legacies over the long term.

During the early Cold War in particular, the CIA played key roles in the overthrow of several political leaders whom the United States feared were too close to the Soviet Union. In Operation TPAJAX in 1953, the CIA overthrew the left-wing Iranian prime minister Mohammed Mossadegh and restored the pro-Western shah to power (Daugherty, 2006, p. 137). The CIA's success came partially from political action—backing anti-Mossadegh elements in the military, clergy, and merchant class—which laid the groundwork for the coups. In the aftermath of the operation, CIA officer in charge Kermitt Roosevelt stated that if "we, the CIA, are ever going to try something like this again, we must be absolutely sure that the people and the army want what we want. If not, you'd better give the jobs to the Marines" (Daugherty,

2006, p. 138). Former director of Central Intelligence Stanfield Turner offered a similar assessment: "The Mossadegh government's political base was weak and susceptible to being toppled. The CIA simply gave it the final push" (Turner, 1985, p. 78).

Similarly, during Operation PBSUCCESS, the CIA helped remove leftwing Guatemalan leader Jacobo Árbenz through military pressure and psychological warfare. On June 18, 1954, the CIA broadcasted false rebel messages that Honduran Colonel Armas, at the helm of a 5,000-man force, was invading the country and heading toward Guatemala City. The radio messages—combined with overt pressure from the United States and limited military actions—proved sufficient to push Árbenz to resign (Turner, 1985, p. 79; Daugherty, 2006, p. 138). Turner noted that "this favorable political outcome required only a small effort, and again the government that was overthrown was so weak that only a little push was needed" (Turner, 1985, p. 79).

The September 11, 1973, overthrow of Salvador Allende's regime in Chile followed a similar pattern. In 1969, the United States received intelligence that the Soviet Union was trying to influence the Chilean election and help the Socialists win power (Daugherty, 2006, p. 171). In response, the United States began a large-scale campaign to prevent Allende's election as president, sponsoring some 726 articles, editorials, and broadcasts (Turner, 1985, p. 80). Despite these efforts, Allende was elected president of Chile in 1970. After increasing tension between Allende and the political right, the Chilean military toppled his government, and Allende died in the process. Ultimately, most of the CIA's campaign against Allende was, in fact, nonlethal: it spent $8 million on election campaign material for anti-Allende parties; $4.3 million on media and propaganda; $900,000 influencing a range of Chilean institutions, from unions to student groups; and $200,000 to support the coup itself (Daugherty, 2006, p. 172; Turner, 1985, pp. 80–81).

All three cases—Iran, Guatemala, and Chile—were only ambiguously successful. While Mossadegh was removed in Iran, the shah was eventually later overthrown—a quarter of a century later—by the radical Shi'ite cleric the Ayatollah Ruhollah Moosavi Khomeini, whose regime still undermines American Middle East policy to this day. Árbenz's ouster in Guatemala led to four decades of military rule that resulted in the deaths hundreds of thousands of Guatemalans (Daugherty, 2006, p. 139). In Chile, a similar story played out. While the CIA was eventually cleared of being directly involved in Allende's death or assisting with military strongman Augusto Pinochet's rise to power, the CIA's Chilean contacts were involved in human rights abuses—albeit over the admonitions of CIA officers (Daugherty, 2006, p. 173).

The U.S. support for guerilla movements has produced, if anything, even more ambiguous outcomes. Perhaps the three most famous examples of this type of political warfare came mostly during the Reagan period: the American support for two rebel factions in the Angolan civil war, for the Contra rebels in Nicaragua, and for the mujahedeen in Afghanistan.

Beginning in January 1975, the Ford administration authorized covert funding to Holden Roberto's Front for the Liberation of Angola (FNLA) and Jonas Savimbi's National Union for Total Independence of Angola (UNITA) in their fight against the Marxist Popular Movement for the Liberation of Angola, which was backed by the Soviet Union and Cuba (Daugherty, 2006, pp. 178–79). Congress banned the assistance in December 1975, only to reinstate it a decade later in 1985 (Daugherty, 2006, p. 180, 210). Despite the fact that the Reagan administration provided approximately $50 million a year in funding and even met with UNITA leadership at the White House, the covert action campaign did not produce decisive results (Daugherty, 2006 p. 210). Rather, Angola became the scene of one of the longest-running civil wars of the twentieth century, and it persisted into the twenty-first century (Daugherty, 2006, p. 180, 210).

American support to the Contra rebels in Nicaragua proved even more controversial. The Carter administration supported these groups, which had formed in opposition to Nicaragua's Marxist Sandinista government, and secured congressional approval for $75 million in aid (Woodward, 1987, p. 111). The Reagan administration expanded these efforts. On December 1, 1981, Ronald Reagan authorized $19 million to the Contras for the purpose of training a 500-man force (Woodward,

1987, p. 589).[5] Ultimately, the program grew to almost $100 million a year, backing thousands of rebels (Daugherty, 2006, p. 204).

As with the previous coups d'états, American support to the Contras—although primarily a military action—included a substantial psychological warfare dimension. The CIA allegedly produced a doctrinal manual on behalf of the Contras (Omang, 1985, p. 5), instructing would-be guerillas to undertake a far-reaching psychological campaign—"the target, then is the minds of the population, all the population: our troops, the enemy and the civilian population" ("Tayacán," 1985, p. 33).

Ultimately, American political warfare was enough to fuel a bloody civil war in Nicaragua but insufficient to overthrow the Sandinista regime. The Sandinistas were voted out of office in the 1990 elections, but only after the Cold War had ended (BBC, 2004). As progress slowed and the casualties mounted, Congress grew increasingly disillusioned with the covert action. In 1982 and then again in 1984, the Democratic-controlled Congress tried to block funding for the Contras through the Boland Amendment. Some in the Reagan administration were unwilling to give up on the program and continued to fund the Contras through covert arms sales to the Islamic Republic of Iran. The scandal ultimately led to the resignation of Reagan's National Security Adviser Admiral John Poindexter and the firing and later prosecution of National Security Council staffer Marine lieutenant colonel Oliver North (PBS, "General Article," n.d.).

The funding of mujahedeen in Afghanistan is yet another example of the law of unintended consequences. After the Soviet Union invaded Afghanistan in 1979 at the request of its left-wing government, the Carter administration began funding the Islamic-inspired rebels known as the mujahedeen. Carter's National Security Adviser Zbigniew Brzezinski gave one reason: "We now have an opportunity of giving to the USSR its Vietnam" (Daugherty, 2006, p. 189). Thanks to well-positioned backers, like Congressmen Charlie Wilson, military support to the mujahedeen grew dramatically under the Reagan administration, from $100 million in 1981 to $120 million in 1983, $250

5 Some sources give a slightly larger $20 million figure (see Daugherty, 2006, p. 204).

million in 1984 to eventually $700 million in 1988 (Daugherty, 2006, p. 206; for the popularized account, see Crile, 2004). All in all, more than $2 billion went to the anti-Soviet rebels, forcing the Soviets to withdraw from Afghanistan, and "it in no small amount contributed to the final downfall of the Soviet Union" (Daugherty, 2006, p. 189).

The holy war against the Soviet Union in Afghanistan, however, also attracted volunteers from across the Arab world, including Osama bin Laden, then a wealthy young Saudi Arabian sheik and the future founder of al Qaeda, who got his first taste of jihad there (Coll, 2004, pp. 155–64). By the end of the conflict, Afghanistan was awash in weapons—by some estimates, it had more antipersonnel weapons in 1992 than India and Pakistan combined (Coll, 2004, p. 238). Worse yet, after the collapse of the Soviet-backed regime, Afghanistan was plunged into a civil war that eventually brought the Islamic fundamentalist Taliban to power and ultimately created a safe haven for bin Laden and terrorism that would pave the way for the 9/11 terrorist attacks.

Economic

Economic tools of political warfare also came of age during the Cold War. Perhaps the most famous use of American economic might to serve political ends was the Marshall Plan. Named for the general turned Secretary of State George Marshall, the initiative started in 1947, with the United States stepping in to bail out Turkey and Greece. In subsequent years, the American aid program expanded. From 1948 to 1951, the United States provided over $13.3 billion to 16 Western European countries to help them recover from the devastation of World War II. The aid served both a humanitarian and a political purpose: By strengthening the economies of Western Europe, the United States believed it could check the spread of Communism. The approach worked: Western Europe recovered and remained firmly democratic. Later on in the Cold War, the United States would use foreign aid as a tool to fight the spread of Communism in Central America, Southeast Asia, and elsewhere (Epstein and Weed, 2009, p. 1).

Ultimately, the Marshall Plan led to more institutionalized efforts to incorporate development into American foreign policy. In 1961, by executive order, President John F. Kennedy created the United States Agency for International Development ("USAID History," 2015). The Kennedy and Johnson administrations saw foreign aid as an important tool in the battle against Communism. After the Cuban revolution, the United States increased American aid to Latin America from $157 million in 1960 to $989 million in 1965 (Lancaster, 2007, p. 69). In the latter half of the 1960s and early 1970s, USAID's priorities changed with America's strategic focus to combat the rise of Communism in Southeast Asia. Thus, South Vietnam, Cambodia, and Laos became prime recipients (Lancaster, 2007, p. 75). Under the Ford and Carter administrations, the United States used foreign aid to incentivize an Egyptian and Israeli ceasefire and peace accord (Lancaster, 2007, pp. 78–79). Even the Reagan administration, which came into office initially skeptical of foreign aid, increased it to Central America from $250 million to $1 billion over the course of the 1980s, largely to help stave off leftist challenges in places like Nicaragua (Lancaster, 2007, p. 80). While the United States used foreign aid for humanitarian purposes as well, foreign aid during the Cold War in particular was seen as a carrot and a method to wield political influence.

On the coercive end, the United States employed economic sanctions to shift politics within the targeted states. It targeted Cuba in 1960 (in response to the Communist takeover) and Iran in 1979 (first in response to the Iran hostage crisis and later in response to the Iranian-backed Hezbollah's bombing of the U.S. Marine barracks in Lebanon) with comprehensive sanctions (Masters, 2015; Laub, 2015). The United States joined multilateral sanctions as well. In two of the rare uses of United Nations sanctions during the Cold War, the Security Council passed sanctions against Rhodesia in response to its unilateral declaration of independence and against South Africa in an effort to force an end to apartheid (Masters, 2015).

The United States employed more covert economic tools against the Soviet Union. In NSDD 54, *United States Policy Towards Eastern Europe*, signed on September 2, 1982, Reagan directed the U.S. government to promote economic liberalization in Eastern Europe as a means to break the Soviet bloc (Daugherty, 2006, p. 198). NSDD 54 was followed by NSDD 66, *Protracted Economic Warfare Against the USSR*, which established "new strategies for increasing economic hardship in the Soviet Union by coopting European governments as partners" (Daugherty, 2006, p. 199). To this end, the United States developed a covert program of sabotage. It sent doctored blueprints, defective equipment, and computer chips with embedded flaws in them—all in an effort to deny the Soviets the use of American technology (Daugherty, 2006, p. 200).

U.S. Political Warfare after the Cold War

After the Cold War ended, the United States found itself without a major ideological rival, and its interest in political warfare waned. While the United States still employed some of the same tools of political warfare—propaganda, aid to opposition groups, economic subversion, and the like—many commentators argue that the United States grew rusty in this domain (for examples, see Jenkins, 2005; Gaffney, 2005). After the collapse of the Soviet Union in 1991, USIA became viewed as a Cold War relic, and Congress abolished the agency in 1999, folding its remaining functions into the State Department (Nakamura and Weed, 2009, pp. 15–16). Foreign aid was also cut in half during this period, from approximately 4.5 percent of discretionary spending in 1984 to a mere 2 percent in 2002 (Tarnoff and Lawson, 2011, p. 17). The 9/11 attacks and the subsequent wars in Iraq and Afghanistan sparked only partial interest in American policy circles in political warfare. In a RAND report published shortly after 9/11, Brian Michael Jenkins argued that political warfare—and influence operations more broadly—needed to be part of any strategy to combat al Qaeda (Jenkins, 2002, p. 24). And yet, the extent to which the military and American policy community embraced the subject remains debatable.

On the information warfare front, while the United States abolished USIA in 1999, Voice of America and RFE/RL continue on. Propaganda and psychological warfare still figure in the War on Terrorism. Some analyses even suggest that the United States should employ arts and literature in ways similar to what was done during the Cold War to help influence the Arab world (Schwartz et al., 2009). The United

States has an active community engagement program as part of its counter-radicalization campaign to try to steer at-risk individuals away from becoming terrorists, although the United States still struggles to counter online terrorist propaganda (see Bjelopera, 2014, pp. 18–19).

The United States has used psychological warfare in a more tactical sense as well. Early on in the Iraq War, the United States released photographs of the disheveled Saddam Hussein soon after he was pulled from his hiding place and later when he underwent a medical examination (Dreazen et al., 2003). The idea was partially to lend veracity to the coalition claims that they had captured him and partially to undermine his support—and with it, the growing insurgency—in Iraq. Later, in 2006, the United States publicized a video of then al Qaeda in Iraq leader Abu Musab al Zarqawi fumbling to fire an M-249 light machine gun, as part of an effort to undermine his authority in the organization. As coalition spokesperson Major General Rick Lynch said, "And his close associates around him . . . do things like grab the hot barrel of the machine gun and burn themselves. It makes you wonder" (Hernandez, 2006).

The United States has taken political action in the post–Cold War age as well. While its efforts may not be as focused as they were under the Cold War, the United States has retained democracy promotion as a stated policy objective (Department of State, n.d.). It has politically backed a series of pro-democracy advocates in authoritarian countries, from Aung San Suu Kyi in Burma to the Dalai Lama in Tibet to democratic reformers in the former Soviet republics in Eastern Europe.

Occasionally, the United States has also employed more militarized forms of political warfare and overtly supported rebel groups. In Iraq, the United States has supported Kurdish groups, first against the Hussein regime and more recently as a proxy against the Islamic State. In Afghanistan, the United States effectively provided military support to the Northern Alliance, leading to the downfall of the Taliban regime. Importantly, though, this same aid proved insufficient to stabilize the country after the regime's collapse, requiring a large conventional American presence that has remained there until today.

The United States has tried to repeat this model elsewhere, with varying degrees of success. It provided about $1 billion in intelligence and air support—as part of a NATO intervention—on behalf of the Libyan rebels fighting the Muammar Gaddafi regime (Barry, 2011). While the support helped the rebels overthrow Gaddafi, this intervention—like its Cold War predecessors—fell victim to unintended consequences, as Libya collapsed into chaos and became a safe haven for the Islamic State (Cruickshank et al., 2014). Even more recently, the United States tried to train Syrian rebels to fight against both the Assad regime and the Islamic State in Iraq. Despite spending approximately $500 million, the program produced only a handful of rebels—most of whom were killed, captured, or defected—and was subsequently terminated (Starr, Kopan, and Acosta, 2015).

Finally, without the same pressing need to stem the rise of Communism, foreign aid has declined dramatically after the Cold War as a percentage of national income (Epstein and Weed, 2009, p. 1). Nonetheless, in the 2002 National Security Strategy of the United States, the Bush administration publically committed to "use our foreign aid to promote freedom and support those who struggle non-violently for it, ensuring that nations moving toward democracy are rewarded for the steps they take" ("The National Security Strategy of the United States of America," 2002, p. 4). While it is debatable how much the Bush administration actually used foreign aid to advance democracy (authoritarian regimes in Egypt, Jordan, and Pakistan remained near the top of the list of recipients through 2008), the Bush administration clearly *intended* to try to use economic power as a means to coerce internal political changes in foreign governments ("Top Ten Foreign Aid Recipients, 2008 and 2012," n.d.). President Barack Obama, likewise, called for placing foreign aid at the heart of his strategy to counter the rise of the Islamic State, arguing that development helps undermine a root cause of terrorism (Kamisar, 2015).

The United States has also embraced the use of sanctions to coerce political behavior, although with mixed success. After the Cold War, the United States—and the world community, through bodies such as the United Nations—turned to sanctions far more frequently as a tool of statecraft to shape an adversary's domestic behavior (see Masters, 2015). The United States—in conjunction with its allies—imposed a series of sanctions against Saddam Hussein's Iraq in an attempt to force him to comply with United Nations weapons inspections after Desert Storm. More recently, the United States has employed multilateral sanctions to pressure Iran to give up its nuclear ambitions and to force Russia to cease its aggression in Ukraine. With the possible exception of the Iran sanctions, these efforts have often failed to produce their desired effects.

Summary

As the foregoing history attests, political warfare is as old and as "American" as the United States itself (Boot and Doran, 2013). While the United States often views itself as the target of political warfare—and, indeed, the other chapters in this volume attempt to identify and guard against potential adversaries' use of political warfare—it is equally important to recognize that the United States has practiced this form of warfare since its founding (see Table 2.1). All major tools of political warfare have deep historical roots in American foreign policy, dating back to the attempts by the American colonists to stir up rebellion in their northern neighbor. Before formulating responses to the political warfare activities of U.S. adversaries and competitors today, therefore, it behooves U.S. strategists to first evaluate the lessons suggested by America's own historical experience with these activities.

It is worthwhile to start with a cautionary, if obvious, point: Political warfare in the twenty-first century is different than it was in the twentieth century. Today's political warfare has to adapt not only to a variety of new technologies—from the Internet to social media—but also to a fundamentally different strategic problem represented by those technologies. Throughout much of the Cold War, the challenge was how to influence closed societies behind the Iron Curtain. At that time, with only a handful of major media outlets, a state could plausibly control the public message. Covert operations could also remain secret, at least in theory.

By contrast, the challenge today is different. With a handful of exceptions such as North Korea, many potential American adversaries are more open than the Soviet Union was, and their citizens have access to more information from a wider range of sources than ever

Table 2.1
The United States and Political Warfare Tactics

Tool	The Early History (Revolution–1945)	Cold War (1946–1991)	Present Day (1992–Present)
Diplomacy/ Political	• Abolition of slavery • Recognition of Czechoslovak nationalism, Jews in Palestine in WWI	• Aid to democratic Western European political parties • Aid to Solidarity	• Recognition of pro-democracy movements and leaders (e.g., Dalai Lama, Aung San Suu Kyi)
Information/ Cyber	• Propaganda to encourage Canada to revolt against Great Britain	• Congress for Cultural Freedom • RFE/RL	• Public diplomacy to the Arab World, post 9/11 • Tactical propaganda to undermine individual leaders (Saddam Hussein, Abu Musab al Zarqawi)
Military	• OSS training of resistance movements in France, Burma, and Thailand	• Coups d'états (Iran, Guatemala, Chile) • Support for the Contras, mujahedeen	• Aid to the Northern Alliance (Afghanistan), Kurds in Iraq, Libyan and Syrian rebels
Economic (Aid, Coercion, Subversion)	• Sanctions against Japan on oil • U.S. Fruit Company's influence on Central American governments	• Marshall Plan • Economic sabotage against Soviet Union	• Economic aid for democracy promotion • Sanctions against Iraq, Iran, and Syria

NOTE: This is an illustrative, rather than comprehensive, list of examples.

before. Russia, China, and other states are using tools to continue to monitor and restrict their citizens' access to information and communications, and the U.S. government has devoted some efforts to defend freedom of speech. However, the United States faces another problem beyond the ability to communicate with citizens in closed societies: developing a message that will resonate among a target audience who can tap into an increasingly wide variety of alternative information sources. Yet another difficulty is that, in an age of ubiquitous communications and dissemination platforms such as Wikileaks, the ability of the United States to conduct operations that remain covert, such as it conducted during the Cold War, is much diminished. While these trends do not wholly invalidate the American experience with political warfare during the Cold War, they should force a recognition that political warfare has changed since Kennan's day.

References—Chapters 1 and 2

Alexiev, Alexander, "Comment on Political Strategies for General Wars: The Case of Eastern Europe," in *Political Warfare and Psychological Operations: Rethinking the US Approach*, Carnes Lord and Frank R. Barnett, eds., Washington, D.C.: National Defense University Press, 1989.

Barnett, Frank R., "Afterword—Twelve Steps to Reviving American PSYOP," in *Political Warfare and Psychological Operations: Rethinking the US Approach*, Carnes Lord and Frank R. Barnett, eds., Washington, D.C.: National Defense University Press, 1989.

Barnett, Frank R., and Carnes Lord, "Introduction," in *Political Warfare and Psychological Operations: Rethinking the US Approach*, Carnes Lord and Frank R. Barnett, eds., Washington, D.C.: National Defense University Press, 1989.

Barry, John, "America's Secret Libya War," *Daily Beast*, August 30, 2011. As of December 7, 2015:
http://www.thedailybeast.com/articles/2011/08/30/america-s-secret-libya-war-u-s-spent-1-billion-on-covert-ops-helping-nato.html

BBC, "Reagan and the 'Iran-Contra' Affair," June 5, 2004. As of December 6, 2015:
http://news.bbc.co.uk/2/hi/americas/269619.stm

Bjelopera, Jerome P., "Countering Violent Extremism in the United States," *Congressional Research Service*, R42553, February 19, 2014. As of December 7, 2015:
https://www.fas.org/sgp/crs/homesec/R42553.pdf

Boot, Max and Michael Doran, "Political Warfare," *Policy Innovation Memorandum No. 33, Council on Foreign Relations*, June 7, 2013:
http://www.cfr.org/wars-and-warfare/political-warfare/p30894

Center for the Study of Intelligence, "Origins of the Congress for Cultural Freedom, 1949–50," *Central Intelligence Agency*, 2008. As of November 24, 2015:
https://www.cia.gov/library/center-for-the-study-of-intelligence/csi-publications/csi-studies/studies/95unclass/Warner.html

Clausewitz, Carl, *On War*, Michael Howard and Peter Paret, ed. and trans., Princeton, N.J.: Princeton University Press, 1984.

Codevilla, Angelo M., "Political Warfare," in *Political Warfare and Psychological Operations: Rethinking the US Approach*, Carnes Lord and Frank R. Barnett, eds. Washington, D.C.: National Defense University Press, 1989.

Cohen, Eliot A., *Conquered into Liberty: Two Centuries along the Great Warpath that Made the American Way of War*, New York: Free Press, 2011.

Coll, Steve, *Ghost Wars: The Secret History of the CIA, Afghanistan and Bin Laden, from the Soviet Invasion to September 10, 2001*, New York: Penguin Books, 2004.

Corke, Sarah-Jane, "George Kennan and the Inauguration of Political Warfare," *The Journal of Conflict Studies*, Vol. 26, No. 1, Summer 2016, pp. 101–20.

Crile, George, *Charlie Wilson's War: The Extraordinary Story of How the Wildest Man in Congress and a Rogue CIA Agent Changed the History of Our Times*, New York: Grove Press, 2004.

Cruickshank, Paul, Nic Robertson, Tim Lister, and Jomana Karadsheh, "ISIS comes to Libya," CNN.com, November 18, 2014. As of December 7, 2015: http://www.cnn.com/2014/11/18/world/isis-libya/

Daugherty, William J., *Executive Secrets: Covert Action & the Presidency*, Lexington, Ky.: University of Kentucky Press, 2006.

Department of State, "Democracy," n.d. As of December 7, 2015: http://www.state.gov/j/drl/democ/

Dreazen, Yochi J., Farnaz Fassihi, Greg Jaffe, and Tom Hamburger, "U.S. Troops Capture Saddam Hussein: Dramatic Seizure Changes Reality from Baghdad to the White House," *Wall Street Journal*, December 15, 2003. As of December 7, 2015:
http://www.wsj.com/articles/SB10714006529014900

Epstein, Susan B., and Matthew C. Weed, "Foreign Aid Reform: Studies and Recommendations," *Congressional Research Service*, R40102, July 28, 2009. As of December 8, 2015:
https://www.fas.org/sgp/crs/row/R40102.pdf

Final Report of the Select Committee to Study Governmental Operations with Respect to Intelligence Activities, United States Senate: Together with Additional, Supplemental, and Separate Views, Washington D.C.: Government Printing Office, April 26, 1976. As of November 30, 2015:
https://archive.org/stream/finalreportofsel01unit/finalreportofsel01unit_djvu.txt

Fischer, Benjamin B., "Solidarity, the CIA and Western Technology," *International Journal of Intelligence and Counterintelligence*, Vol. 25, No. 3, December 2012, pp. 427–69.

Gaffney, Frank R., Jr., "The Political Front: Information is an Important Tool on the Road to Victory," *National Review Online*, December 7, 2005. As of November 30, 2015:
http://www.nationalreview.com/article/216186/political-front-frank-j-gaffney-jr

Galula, David, *Counterinsurgency Warfare: Theory and Practice*, New York: Fredrick Praeger, 1964.

Godson, Roy, *Dirty Tricks or Trump Cards: U.S. Covert Action & Counterintelligence*, New Brunswick, N.J.: Transaction Publishers, 2008.

Hernandez, Nelson, "Now Playing in Iraq: Zarqawi Outtakes," *Washington Post*, May 5, 2006. As of December 7, 2015:
http://www.washingtonpost.com/wp-dyn/content/article/2006/05/04/AR2006050400454.html

Hoffman, Frank, "On Not-So-New Warfare: Political Warfare vs. Hybrid Threats," *War on the Rocks*, July 28, 2014. As of November 17, 2015:
http://warontherocks.com/2014/07/on-not-so-new-warfare-political-warfare-vs-hybrid-threats/

"Jedburghs," n.d. As of January 21, 2016:
http://www.soc.mil/OSS/jedburghs.html

Jenkins, Brian Michael, *Countering Al Qaeda: An Appreciation of the Situation and Suggestions for Strategy*, Santa Monica, Calif.: RAND Corporation, MG-1620, 2002. As of November 30, 2015:
https://www.rand.org/pubs/monograph_reports/MR1620.html

———, "Strategy: Political Warfare Neglected," *San Diego Union-Tribune*, June 26, 2005. As of November 30, 2015:
http://www.rand.org/blog/2005/06/strategy-political-warfare-neglected.html

Kamisar, Ben, "Obama: Foreign Aid Key Part of ISIS Strategy," *Hill*, March 16, 2015. As of December 9, 2015:
http://thehill.com/blogs/blog-briefing-room/news/235877-obama-foreign-aid-necessary-part-of-isis-strategy

Kennan, George, to the Secretary of State, "861.00/2-2246: Telegram," February 22, 1946. As of November 30, 2015:
http://nsarchive.gwu.edu/coldwar/documents/episode-1/kennan.htm

Lancaster, Carol, *Foreign Aid: Diplomacy, Development, Domestic Politics*, Chicago: The University of Chicago Press, 2007.

Laub, Zachary, "International Sanctions on Iran," *CFR Backgrounders*, July 15, 2015. As of December 9, 2015:
http://www.cfr.org/iran/international-sanctions-iran/p20258

Lord, Carnes, "The Psychological Dimension in National Strategy," in *Political Warfare and Psychological Operations: Rethinking the US Approach*, Carnes Lord and Frank R. Barnett, eds., Washington, D.C.: National Defense University Press, 1989.

Lucas, Scott, and Kaeten Mistry, "Illusions of Coherence: George F. Kennan, U.S. Strategy and Political Warfare in Early Cold War 1946–1950," *Diplomatic History*, Vol. 33, No. 1, January 2009, pp. 39–66.

Lynn, Katalin Kádár, "At War While at Peace: United States Cold War Policy and the National Committee for a Free Europe, Inc.," in *The Inauguration of Organized Political Warfare*, Katalin Kádár Lynn, ed. Saint Helena, Calif.: Helena History Press, 2013.

Masters, Jonathan, "What Are Economic Sanctions?," *CFR Backgrounder*, April 8, 2015. As of December 9, 2015:
http://www.cfr.org/sanctions/economic-sanctions/p36259

Mistry, Kaeten, "The Case for Political Warfare: Strategy, Organization and US Involvement in the 1948 Italian Election," *Cold War History*, Vol. 6, No. 3, 2006, pp. 301–29.

Nakamura, Kennon H., and Matthew C. Weed, "U.S. Public Diplomacy: Background and Current Issues," *United States Information Agency*, R-40989, December 18, 2009. As of January 18, 2016:
https://www.fas.org/sgp/crs/row/R40989.pdf

"The National Security Strategy of the United States of America," 2002. As of December 8, 2015:
http://www.dtic.mil/dtic/tr/fulltext/u2/a407178.pdf

Needell, Allan, "'Truth Is Our Weapon': Project TROY, Political Warfare, and Government-Academic Relations in the National Security State," *Diplomatic History*, Vol. 17, No. 3, 1993, pp. 399–420.

"NSC 68: United States Objectives and Programs for National Security," n.d. [1950]. As of November 30, 2015:
http://fas.org/irp/offdocs/nsc-hst/nsc-68-9.htm

Office of the Historian, "Japan, China, the United States and the Road to Pearl Harbor, 1937–41," *Department of State*. As of December 13, 2015:
https://history.state.gov/milestones/1937-1945/pearl-harbor

"The Office of Strategic Services—America's First Intelligence Agency," Central Intelligence Agency, November 21, 2012. As of December 14, 2015:
https://www.cia.gov/about-cia/cia-museum/experience-the-collection/text-version/stories/the-office-of-strategic-services-americas-first-intelligence-agency.html

Omang, Joanne, "A Historical Background to the CIA's Nicaragua Manual," in *Psychological Operations in Guerilla Warfare: The CIA's Nicaragua Manual*, New York: Vintage Books, 1985.

"On the Inauguration of Organized Political Warfare," *Policy Planning Staff Memorandum* 269, Washington D.C., May 4, 1948. As of November 10, 2015:
http://academic.brooklyn.cuny.edu/history/johnson/65ciafounding3.htm

Paddock, Alfred H., Jr., "Military Psychological Operations," in *Political Warfare and Psychological Operations: Rethinking the US Approach*, Carnes Lord and Frank R. Barnett, eds., Washington, D.C.: National Defense University Press, 1989.

———, *U.S. Army Special War: Its Origins*. Lawrence, Kan.: University of Kansas Press, 2002.

PBS, "General Article: The Iran-Contra Affair," *American Experience*, n.d. As of December 6, 2015:
http://www.pbs.org/wgbh/americanexperience/features/general-article/reagan-iran/

———, "Proclamation of Earl of Dunmore, 1775," n.d. As of December 2, 2015:
http://www.pbs.org/wgbh/aia/part2/2h42.html

Reagan, Ronald, "Address to the National Association of Evangelicals ('Evil Empire Speech')," March 8, 1983. As of December 1, 2015:
http://voicesofdemocracy.umd.edu/reagan-evil-empire-speech-text/

Rowen, Henry S., "Political Strategies for General Wars: The Case of Eastern Europe," in *Political Warfare and Psychological Operations: Rethinking the US Approach*, Carnes Lord and Frank R. Barnett, eds., Washington, D.C.: National Defense University Press, 1989.

Saunders, Frances Stonor, *The Cultural Cold War: The CIA and the World of Arts and Letters*, New York: The New Press, 2013.

Schwartz, Lowell H., Todd C. Helmus, Dalia Dassa Kaye, and Nadia Oweidat, *Barriers to the Broad Dissemination of Creative Works in the Arab World*, Santa Monica, Calif.: RAND Corporation, MG-879, 2009. As of December 9, 2015:
http://www.rand.org/pubs/monographs/MG879.html

Smith, Paul A., Jr., *On Political War*, Washington, D.C.: National Defense University Press, 1989.

Speier, Hans, "The Future of Psychological Warfare," *Public Opinion Quarterly*, Spring 1948, pp. 5–18.

Starr, Barbara, Tal Kopan, and Jim Acosta, "U.S. Suspending Program to Train and Equip Syrian Rebels," CNN.com, October 9, 2015. As of December 7, 2015:
http://www.cnn.com/2015/10/09/politics/us-syria-rebels-arms-program-suspended/

Tarnoff, Curt, and Marian Leonardo Lawson, "Foreign Aid: An Introduction to U.S. Programs and Policy," *Congressional Research Service*, R-40213, February 10, 2011. As of January 18, 2016:
https://www.fas.org/sgp/crs/row/R40213.pdf

"Tayacán," *Psychological Operations in Guerilla Warfare: The CIA's Nicaragua Manual*, New York: Vintage Books, 1985.

"Top Ten Foreign Aid Recipients, 2008 and 2012," n.d., *ABC News*. As of December 8, 2015:
http://abcnews.go.com/Politics/OTUS/fullpage/top-10-us-foreign-aid-recipients-17534761

"Transcript of Articles of Confederation (1777)." As of December 14, 2015:
http://www.ourdocuments.gov/doc.php?flash=true&doc=3&page=transcript

Trinquier, Roger, *Modern Warfare: A French View of Counterinsurgency*, Daniel Lee, trans., London: Pall Mall Press, 1964. As of November 30, 2015:
http://usacac.army.mil/cac2/cgsc/carl/download/csipubs/modernwarfare.pdf

Turner, Stanfield, *Secrecy and Democracy: The CIA in Transition*, Boston: Houghton Mifflin Company, 1985.

United States Army Special Operations Command, "SOF Support to Political Warfare White Paper," March 10, 2015. As of November 30, 2015:
https://dl.dropboxusercontent.com/u/6891151/Support%20to%20Political%20Warfare%20White%20Paper%20v2.3-RMT%20%2810MAR2015%29%20%20%20.pdf

"USAID History," May 28, 2015. As of January 18, 2016:
https://www.usaid.gov/who-we-are/usaid-history

Vereté, Mayir, "The Balfour Declaration and Its Makers," *Middle Eastern Studies*, Vol. 6, No. 1, January 1970, pp. 48–76.

"VOA History," n.d. As of January 18, 2016:
http://www.insidevoa.com/section/voa_history/2330.html

"What Was OSS?" *Central Intelligence Agency*, June 28, 2008. As of December 14, 2015:
https://www.cia.gov/library/publications/intelligence-history/oss/art03.htm

Woodward, Bob, *Veil: The Secret Wars of the CIA 1981–1987*, New York: Pocket Books, 1987.

CHAPTER THREE

Case Study: Russia

This chapter examines Russia's understanding, practice, and potential for using political warfare in the future. Russia has repeatedly used proxies, propaganda, and covert action, among other tactics, to undermine European security, most recently in Ukraine. There is good reason to expect that Russia will continue to use these tactics in the future in ways that will threaten core U.S. interests in Europe and around the world.

To understand the nature, limits, and capabilities of Russian political warfare, this chapter proceeds in three parts. First, we summarize the political and academic discourse around political warfare in Russia, highlighting the growth of the Russian concept of "new generation warfare," which sees nonmilitary elements playing an increasing role in warfare. We trace the origins of this concept to Russian interpretations of U.S. activities, Soviet and Russian foreign policy, and domestic political trends in Russia. Second, we catalog and characterize Russian capabilities to undertake different elements of political warfare. We categorize these elements as diplomatic and proxy, information, cyber, military, intelligence, and economic. We also discuss the development of coordination across the Russian government. Finally, we use a case study on Russian political warfare in Estonia to develop new insights into how Russia undertakes political warfare in practice. The case study emphasizes that Russia tends not to fully orchestrate political warfare from beginning to end, but instead acts to increase tension to make confrontation more likely, and then takes advantage of crises when they arise.

This chapter draws on a number of sources, including official documents, commentary and analysis in the Russian and Western policy and academic literature, interviews and discussions with subject matter experts and officials, and the recent use of political warfare in countries including Estonia, Georgia, Ukraine, and Moldova.

Russian Understanding of Political Warfare

In the past few years, a significant development within the Russian military establishment has been the new concept of "new generation warfare," which roughly aligns with Western understandings of political warfare. U.S. analysts have emphasized that Russian activities in Ukraine appear to reflect this new line of thinking,[1] and point to an increased threat of Russian political warfare.[2] Russian officials and analysts see "new generation warfare" as a description of how warfare has evolved in general, implying an intent both to understand the threat to Russia and also to adapt and utilize the Russian state to achieve its political objectives in the future.

The essence of "new generation warfare" is the growing use of nonmilitary means in combination with military tools to achieve political ends.[3] One of the clearest and most commonly cited depictions of the changing nature of warfare comes from the Russian Chief of

---

[1] An account of the Ukraine crisis published by USASOC notes, "Gerasimov's model for modern conflict is a theoretical adaptation of emerging ideas about warfare, but elements of these ideas clearly pertain to Russia's unconventional warfare in Ukraine and Crimea in 2013–2014" and directly links the theoretical model to events in Ukraine. ("'Little Green Men': A Primer on Modern Russian Unconventional Warfare," n.d.; Dickey et al., 2015, p. 221).

[2] A USASOC white paper notes, "Russian UW is thus the central, most game-changing component of a Hybrid Warfare effort involving conventional forces, economic intimidation of regional countries, influence operations, force-posturing all along NATO borders, and diplomatic intervention. . . . Russia's current military operations in Ukraine and recent operations in the region provide examples of Russia's New Generation Warfare" (USASOC, 2014, p. 4, 29).

[3] Dima Adamsky writes, "To Gerasimov and other experts, NGW [new generation warfare] is an amalgamation of hard and soft power across various domains, through skillful

the General Staff, Valery Gerasimov, in a February 2013 article, "The Value of Science in Prediction," in the *Military Industrial Courier*:

> In the 21st century we have seen a tendency toward blurring the lines between the states of war and peace. Wars are no longer declared and, having begun, proceed according to an unfamiliar template.
>
> The role of non-military means of achieving political and strategic goals has grown, and, in many cases, they have exceeded the power of force of weapons in their effectiveness. The focus of applied methods of conflict has altered in the direction of the broad use of political, economic, informational, humanitarian, and other non-military measures—applied in coordination with the protest potential of the population.
>
> All this is supplemented by military means of a concealed character, including carrying out actions of informational conflict and the actions of special operations forces. The open use of forces—often under the guise of peacekeeping and crisis regulation—is resorted to only at a certain stage, primarily for the achievement of final success in the conflict. (Gerasimov, 2013)

Gerasimov draws from extensive discussion in Russian military journals about the changing nature of warfare. In 2003, the president of the Russian Academy of Military Sciences, Makhmut Gareyev, wrote, "The point is that in recent decades we have become witnesses to how entire nations and coalitions of nations have come to be destroyed in the course of confrontation in the international arena without the direct use of armed force. . . . The correlation of political, diplomatic, economic, information, psychological, and military means of fighting in the international arena have changed markedly in contemporary times. The significance and proportionate share of nonmilitary means have increased significantly" (Gareyev, 2003). In 2010, Sergei Chekinov and Sergei Bogdanov similarly observed that improving technology

---

application of coordinated military, diplomatic, and economic tools" (Adamsky, 2015, p. 23; Berzins, 2014).

has made conventional weapons more effective and full-scale war more catastrophic, implying that "the role and significance of [nonmilitary measures of interstate confrontation] continues to rise" (Chekinov and Bogdanov, 2010).

A 2013 article by the same authors summarizes the developing literature on new generation warfare. They highlight how technological advances make it more possible for countries to undermine the political and military leadership of their adversaries, both through information operations (including not only propaganda but also bribery, cyber attacks, etc.) and high-precision long-range strikes. Analysts have used their article to identify discrete stages of a new generation war, implying a concrete plan (Berzins, 2014, p. 6). Instead, their writing emphasizes the new possibilities emerging from technological and social trends, rather than a specific plan of attack. Before the beginning of a war, for example, "heavy propaganda" may be used to "depress the moral and psychological feelings of the civilian population and armed forces." During the opening period of the war, electronic warfare, high-precision weaponry, and other kinetic means could be used to "destroy critical government and military command centers," and thereby undermine the basic system of government. During the closing period, ground forces (which they note are a critical component of new generation warfare), will "roll over the remaining points of resistance" (Chekinov and Bogdanov, 2010, p. 22). Though the effectiveness of new generation warfare involves the combination of military and nonmilitary measures, nonmilitary measures are believed to be more important; indeed, the literature suggests an ideal ratio of 4 to 1 for nonmilitary to military measures (Adamsky, 2015, p. 23).

"Information warfare" occupies a central position in this discussion, and in the Russian context includes a wide range of activities designed to influence the gathering of information, cognition, decisionmaking, coordination, and the transfer of information. Dima Adamsky writes, "Since, according to [new generation warfare], the main battlefield is consciousness, perception, and strategic calculus of the adversary, the main operational tool is information struggle, aimed at imposing one's strategic will on the other side" (Adamsky, 2015, p. 26). Similarly, Chekinov and Bogdanov observe that "no goal will be achieved in future wars unless one belligerent gains information superiority over the other" (Chekinov and Bogdanov, 2010, p. 13).[4]

Discussion of information warfare is also linked to a belief that Russia is engaged in a civilizational struggle with the West.[5] Russia sees itself as a great power with a rich culture and as a conservative bulwark in the face of encroaching liberalization (Van Herpen, 2016, pp. 28–29). Some within Russia see the Western media, support for civil society, and other efforts intended to support democracy and free exchange as instead intended to undermine the Russian state.[6] In this context, part of the responsibility of the state is seen as defending the information space of its citizens—i.e., their ideas and perspectives—from foreign influence. Defense of Russia's information sphere, from this point of view, also involves control of the information reaching its near-abroad, as well as conducting offensive information warfare in the West to counter Western dominance of information (Darczewska, 2015, p. 30).

Russia's military and civilian strategy documents also emphasize the changing threat to Russia of combined nonmilitary and military means, and the need to respond in kind. The December 2015 National Security Strategy observes that Russia faces "opposition from the United States and its allies," who are containing Russia through "the

4 Russian military writing has examined information warfare for decades, for example identifying the concept of "reflexive control," which involves leading "the adversary to act according to a false picture of reality in a predictable way" (Adamsky, 2015, pp. 26–27; Berzins, 2014).

5 E.g., the Foreign Policy Concept notes, "For the first time in modern history, global competition takes place on a civilizational level, whereby various values and models of development based on the universal principles of democracy and market economy start to clash and compete against each other. Cultural and civilizational diversity of the world becomes more and more manifest." Timothy Thomas writes, "Some still blame the fall of communism on what many Russians term the 'information-psychological' assault from the West sometimes referred to as the so-called Third World War" (Thomas, 2014, p. 123).

6 Putin, for example, explained in July 2012: "Russia's image abroad is formed not by us and as a result is often distorted and does not reflect the real situation in our country or Russia's contribution to global civilization, science and culture. . . . Our country's policies often suffer from a one-sided portrayal. . . . But our fault lies in our failure to adequately explain our position" (Van Herpen, 2016, p. 29).

exertion of political, economic, military, and informational pressure on it." To respond, the strategy notes that Russia is developing and implementing "[i]nterrelated political, military, military-technical, diplomatic, economic, informational, and other measures . . . to ensure strategic deterrence and the prevention of armed conflicts" ("On the Russian Federation's National Security Strategy," 2015, para. 36). Similarly, the 2014 Military Doctrine notes that the characteristics of current military contingencies include the "integrated employment of military force and political, economic, informational or other nonmilitary measures implemented with a wide use of the protest potential of the population and of special operations forces" ("Russian Military Doctrine 2014," 2014, para. 15a). Russia also has a stated doctrine on "information security," which discusses the imperative of protecting the Russian information sphere from the involvement of foreign actors.[7] The information security doctrine identifies a threat from coordinated "information weapons" from foreign entities, noting, for example, the threat of "the ousting of Russian news agencies and media from the national information market, and an increase in dependence on the spiritual, economic, and political areas of public life in Russia on foreign information entities."[8]

Russian strategic documents, as well as analysts of Russian behavior, also use the term "soft power" to understand the threat and use of

---

[7] The doctrine also identifies information activities aimed at "influencing the Russian population in order to sabotage its historical, spiritual and patriotic traditions," as well as at the "incitement of cross-ethnic and social tensions, extremism, and ethnic and religious animosity" among the primary threats to the state's stability. The doctrine emphasizes the Internet as a potential tool of "influence in the global information space, able to penetrate to the adversary's entire territory" (Military Doctrine of the Russian Federation, 2014).

[8] Information Security Doctrine of the Russian Federation (2008). Similarly, the Concept of the Foreign Policy of the Russian Federation, (2013) notes, "Russia will take necessary measures to ensure national and international information security, prevent political, economic and social threats to the state's security that emerge in information space in order to combat terrorism and other criminal threats in the area of application of information and communication technologies, prevent them from being used for military and political purposes that run counter to international law, including actions aimed at interference in the internal affairs and constituting a threat to international peace, security and stability."

integrated nonmilitary tactics to achieve political ends.[9] The 2013 Foreign Policy Concept defines "soft power" as a "comprehensive toolkit for achieving foreign policy objectives building on civil society potential, information, cultural and other methods and technologies alternative to traditional diplomacy." It sees soft power as "an indispensable component of modern international relations" ("Russian Foreign Policy Concept," 2013) and observes the threat, presumably from the West, of "destructive and unlawful use of 'soft power' and human rights concepts to exert political pressure on sovereign states . . . including under the pretext of financing cultural and human rights projects abroad" ("Russian Foreign Policy Concept," 2013). Putin also offered a speech in 2012 highlighting the importance of strengthening Russia's soft power techniques, and some analysts prefer the term "soft power" to "political warfare" as a rubric to understand Russian activities (Putin, 2012; Sergunin and Karabeshkin, 2015, p. 352; Van Herpen, 2016, p. 27).

Beyond the official perspective, more extreme views about the need to engage in political warfare can be found, especially from scholars of "geopolitics." Igor Panarin, dean of the academy for Russia's Foreign Ministry, attributes the widespread recent political changes in the Arab Spring, color revolutions, and Maidan protests to "social control technology and information aggression" from the United States.[10] Aleksandr Dugin, a former professor at the Moscow State University, draws from the term "net-centric warfare" to identify a Western campaign of "netwar," meaning "artificial processes plotted in the West aimed at destabilizing entire regions in the post-Soviet area" (Darczewska, 2014,

---

9 Martin Van Herpen also describes a three-part Russian strategy of "soft power": "mimesis," which involves copying Western methods of influence; "'rollback," "a logical consequence of the Kremlin's vision that the soft power game in an integral part of a zero-sum hard-power game . . . means curtailing, opposing, and possibly forbidding the activities of Western soft-power in states inside Russia;" and "invention," in which Russia uses the open character of Western societies and past KGB-style techniques to enhance Russia's influence abroad (Van Herpen, 2016, pp. 33–34).

10 "The directors of the information war against the USSR (A. Dulles, G. Kennan, D. Rockefeller, H. Kissinger, Z. Brzezinski, R. Reagan) were able to implement a strategic operation in order to promote the ascension on the USSR of the Trotskyites-globalists N.S. Khrushchev and M.S. Gorbachev." Hence, the U.S. led a conspiracy to achieve the fall of the Soviet Union and subsequent weakness of the Russian state (Van Herpen, 2016, p. 8).

p. 17). To counter the West's efforts, both Dugin and Panarin urge Russia to adopt a more organized, centralized, and concerted means for countering Western information war, and they call for unity and coordination among Russia's intelligence services, government, intelligentsia, and all other social strata. They see justification for this approach in Russia's historical mission to protect the global good and to contain the United States' global encroachment (Darczewska, 2014, pp. 14–18, 23–32). Though Dugin and Panarin's views do seem more influential than one might expect given their extreme positions,[11] their influence on policy is probably limited. Dugin, for example, was removed from his position at Moscow State University after calling for the killing of Ukrainian nationalists (Fitzpatrick, 2014; Laruelle, 2012).

The ideas outlined above offer a framework for understanding how Russian officials and analysts understand the growing use of nonmilitary tactics, in coordination with conventional military tools, to achieve political objectives. In many cases, Russian analysis on these subjects may be disconnected from or more advanced than its actual practice. Whatever the practical limitations of Russian capabilities, it does appear that military and civilian officials are increasingly trying to develop Russia's ability to conduct effective political warfare.

Sources of Russian Views of Political Warfare

Russian analysis of the changing nature of warfare and interpretations of political warfare draw from at least three sources. First, Russian observers have carefully analyzed U.S. activities for signs of the changing nature of warfare and the potential Western threat to Russia. For example, Russian officials highlight the threat of U.S.-backed color revolutions, which in their view demonstrate how the United States has utilized nonmilitary tactics, sometimes backed by military force, to achieve its objectives of regime change.[12] Dmitry Gorenberg paraphrases

---

[11] Dugin and Panarin are commonly cited as important thought leaders on information warfare (Darczewska, 2014, pp. 14–18; U.S. Army Special Operations Command, n.d., pp. 15–17).

[12] One article noted, "In the area of politics, the military actions carried out by the United States have asserted its right to use force against any state, and so launched the shaping of a new world order. The role of the United Nations and its Security Council in addressing major

Gerasimov, arguing that "the U.S. can't deal with more equal relations among states, so it is using new tactics to assure its supremacy . . . the United States has developed a new method of warfare, beginning with using non-military tactics to change opposing governments through colored revolutions that utilize the protest potential of the population to engineer peaceful regime change" (Gorenburg, 2014).[13] Similarly, Yuri Baluyevskiy, former Chief of the General Staff, writes:

> For the first time in the history of warfare, relying on information as the fundamental element in the conduct of armed conflict, the collective West managed to imbed in the consciousness of the population of the majority of countries around the world the idea that the United States had the exclusive right to world governance. . . . In Europe, NATO has actively expanded to the east, drawing into its orbit of influence CIS republics. The reconstruction of the consciousness of the population and the

---

global problems has been irreparably undermined. It may be assumed that *military operations like Operation Iraqi Freedom will continue.* Along with the three previous operations, it is merely an intermediate link in the chain of planned forcible actions whose aim is to bring the U.S. closer to achieving world domination" (Tashlykov, 2010, p. 67 [emphasis in original]).

13 In interviews with foreign media, a former head of the Russian domestic security agency (FSB) Nikolai Patrushev similarly described Russian motives in Ukraine: "For the last quarter of a century, these actions were designed to wrest Ukraine and other former Soviet republics away from Russia and to redesign the post-Soviet space in America's interests. The US created the conditions and pretexts for the coloured revolutions and financed them lavishly. . . . Victoria Nuland, the US assistant secretary of state for European and Eurasian affairs, has said that her country spent $5bn between 1991 and 2013 "supporting the aspirations of the people of Ukraine for a stronger, more democratic government." Even open sources, such as congressional documents, show that government spent no less than $2.4bn on various American "aid" programmes to Ukraine between 2001 and 2012. . . . The coup in Kiev, which was clearly carried out with the help of the US, conformed to a classical model worked out in Latin America, Africa and the Middle East. But never before had such a coup impinged on Russia's interests so directly" (Egorov and Patrushev, 2014). A recent article assessing threats to Russia similarly notes, "The situation is complicated by the fact that the United States and its allies, above all those in NATO, take advantage of controlled chaos technologies and actively intervene in all spheres of vital activities, not even drawing the line at covert use of force. Such actions broaden the spectrum of threats and become the main method of achieving military (and in the final account political) goals by nonmilitary means" (Podgornykh and Dolgov, 2014).

> targeted formation of a pseudo-elite in these states allowed the West, at relatively little financial and material cost, to obtain loyal allies on our borders that are more than happy to kill their former compatriots. The apotheosis of this was the Georgian attack on South Ossetia in August of 2008. (Baluyevskiy and Khamzatov, 2014)

While this view of Western tactics may appear paranoid to Western audiences, Adamsky emphasizes that the Kremlin and those around it "genuinely perceive Russia as operating under a long-lasting encirclement which aims to undermine and ultimately destroy it in geopolitical terms. Its current behavior, in its eyes, is a defensive counter attack following a Western aggression across various domains—in international, military, economic-energy, and internal affairs."[14]

Some analysts have noted that Gerasimov's discussion appears to describe Russia's actions in the Crimea and eastern Ukraine and, implicitly, future Russian activity. Robert Coalson (2014), for example, sees Russia putting in place tactics implied by Gerasimov's writings in Ukraine: "it became clear very quickly that Russian politicians, journalists, purportedly nongovernmental organizations, state companies, think tanks, the military, the courts, government agencies and the Duma were all working from the same instructions for the same goals."[15] Other analysts claim Gerasimov is describing U.S. activities, rather than expressing any intention for how Russian forces will develop or behave. Samuel Charap writes, for example, "Gerasimov is

---

[14] Adamsky goes on to note, "With few exceptions, Western experts utilize the term [hybrid warfare] to describe Russian military theory and practice, particularly regarding the crisis in Ukraine and the potential future standoff on its European periphery—especially in the Baltic area. This categorization may be inaccurate. The current Russian thinking and waging of war is different from HW, as perceived in the West. . . . Before the 2014 events in Ukraine, the term appeared in the professional discourse either in reference to the U.S. Threat perception or to categorize one of the recent trends in the U.S. Way of war" (Adamsky, 2015, pp. 21–23).

[15] Coalson further observes, "The view of global affairs presented in this article, I think, accurately reflects an important strain of Kremlin thinking. After all, it is presented in an obscure publication and its exposure to foreign audiences could not have been foreseen" (Coalson, 2014; Galeotti, 2014a).

actually describing what he sees as the new US way of war, not Russian doctrine" (Charap, 2015, p, 53). Charap also notes the inconsistent success of Russian irregular warfare in Ukraine, as well as the fact that Russia needed to use conventional warfare to prevent a separatist defeat in eastern Ukraine in August 2014, as evidence against the likely use of irregular or proxy warfare by Russia in the future. But inconsistent success does not necessarily mean that Russia did not attempt to adopt lessons from new generation warfare into its activities in Ukraine (both in Crimea and eastern Ukraine) or that it will not continue to adapt its forces with new generation warfare in mind in the future.

Second, Russia's thinking about political warfare draws from past Russian and Soviet foreign policy.[16] Russian analyst Fyodor Lukyanov, for example, summarizes Russia's soft power strategy as a plan "to revive Soviet-era practices that were quite effective" ("Why Russia's Soft Power Is Too Soft," n.d.). Polish analyst Jolanta Darczewska argues that Russia's adoption of information warfare concepts represents an attachment "to old methods (sabotage, diversionary tactics, disinformation, state terror, manipulation, aggressive propaganda, exploiting the potential for protest among the local population)" (Darczewska, 2014, p. 7). Archived KGB documents depict exactly the measures that Russian ideologues, politicians, and doctrinal documents attribute to the Western actors ("Actions to Promote Discord," 2007). For example, these documents describe KGB's efforts to destabilize Turkey in the 1950s and draw cleavages between Turkey and the United States using deception, fabricated letters to Turkish officials, strategic content placement in Turkish newspapers, and creation of false opposition groups. The documents further describe means of manipulation of public opinion abroad ("On Human Rights," 2007), use of psychological operations ("KGB Practices," 2007), and other approaches to undermining the power and credibility of individuals, organizations, and countries that stood in opposition to the

---

16 Several Western analysts of Russia emphasized that there was nothing new in the adoption of political warfare tactics by Russia (discussions with U.S. and European analysts, November 2015–February 2016; Russian Measures Short of War Conference, 2016; Wilson, 2005, Chapter One).

Soviet ideology (see, e.g., "Association of the United Postwar Immigrants," 2007). Analysts point to Soviet-era terminology such as "active measures" and *maskirovka* (deception) to explain Russia's behavior (Berzins, 2014, p. 7; Heickerö, 2010, pp. 21-23; Blank, 2013, pp. 31–44).

Third, Russian political warfare draws from its own domestic politics. Russia's political system is perhaps best understood as "competitive authoritarianism," in which there is some form of democratic competition, but the regime is able to unfairly skew the vote in its favor (Levitsky and Way, 2010). The techniques of political manipulation are often referred to as "political technologies," and in general the term "technology" is used to describe a technique of nefarious or unfair manipulation. Common examples of political technologies include covert support for nongovernmental organizations, the creation of fake organizations as a means of infiltration or redirection of sentiment, manipulation of media, and misuse of the justice system to attack opponents ("legal technology"). Peter Pomerantsev writes that Vladislav Surkov, "the political technologist of all Rus"[17] and a senior advisor to Putin, "directed Russian society like one great reality show. He claps once and a new political party appears. He claps again and creates Nashi, the Russian equivalent of the Hitler Youth" (Pomerantsev, 2015, p. 65). In Pomerantsev's account, Surkov and others have a philosophy of government in which "everything is PR"—the regime has no connection to any particular ideology, and instead seamlessly shifts between narratives and controls opposition groups voicing apparently contradictory messages (Pomerantsev, 2015, pp. 65–74). The use of political technology appears fundamental to the regime—indeed, Andrew Wilson describes Russia under Putin as a "virtual democracy," in which "politics should only exist as a series of designer projects, rather than as a real pattern of representation and accountability" (Wilson, 2005, p. 39).[18]

---

17 I.e., the Russian ethnic group.

18 Putin's strategy for governing Russia is also described in a strategy document published in *Kommersant* on May 5, 2000; for an English translation, see Podkopaev, Dawisha, and Nealy, n.d. See also Dawisha, 2014.

It is natural for Putin's regime to draw on these techniques to create political change abroad. Indeed, Wilson highlights how "political technology" used in Russia has been exported and used in the rest of the post-Soviet world,[19] including media manipulation, the setup of manufactured groups to disrupt opponents' campaigns,[20] the creation of fake political parties to distract from the real candidates, and "black PR" or "Kompromat"—"the use of compromising material, real or imagined (financial details, lurid videos, simple slander), against opponents" (Wilson, 2005, p. 47).

Not all of these tactics can be used in an unlimited way in the post-Soviet world. In Ukraine, for example, Wilson observes, "The opportunities for direct ballot stuffing were limited, so the authorities used 'legal technologies' to prevent the opposition [from] registering in many localities and [set] up fake 'clone' branches instead" (Wilson, 2011). In the Baltic states, though Wilson observes "covert action, subterranean party finance and the persistent use of kompromat," he emphasizes that "wholly fake parties don't really exist in the Baltic States: they are too small and their media markets are too free" (Wilson, 2011). Wilson claims that political technology depends on "a powerful but amoral elite; a passive electorate; a culture of information control; and the lack of an external counterpoint, i.e., foreign intervention" (Wilson, 2005, p. 41). Similarly, Steve Levitsky and Lucan Way theorize that competitive authoritarianism can persist only in the absence of Western leverage and of local linkages to Western countries, implying limits on Russia's ability to influence many Eastern European countries (Levitsky and Way, 2010, Chapter Two).

---

[19] Wilson notes, "Russia doesn't have many export industries. . . . Now Russia has political technology. . . . Russian political technologists were not just in Ukraine and elsewhere to make money, although there was plenty of that around. According to Volodymyr Polokhalo, 'they aren't just making money, but serving as Russian agents of influence, "distributors" of Russian interests'" (Wilson, 2005, pp. 57–58).

[20] Examples include "organized evening canvassing by company employees claiming to work for opponents, issued invitation to opponents' rallies in bad weather or rallies that simply failed to happen, handed out free but shoddy goods from opponents" (Wilson, 2005, p. 66).

Thus, there is a rich and evolving discourse within the Russian military and foreign policy establishment on political warfare. Senior leaders take seriously the growing importance of nonmilitary tactics for achieving political objectives. In addition to drawing from views of the changing nature of warfare, Russian thinking about political warfare takes into account long-standing analysis of information warfare, Soviet-era tactics of political warfare, and the use of political-warfare-style tactics in Russian domestic politics. Nevertheless, there may be a substantial gap between the way that Russia thinks about political warfare and the way it actually practices these tactics. Hence, the next section examines in detail the state establishment that actually performs political warfare.

Russian Organizations That Pursue "Political Warfare"

This section surveys the elements of Russian state power that are used to achieve its objectives without the use of military force. The section emphasizes the covert, nefarious, and deniable tools, rather than the overt and official ones. Nonmilitary tactics are also used in kinetic military operations, but the use of these tactics in such operations has a different character than political warfare.

Western officials and analysts have highlighted Russia's capabilities to use political warfare to achieve their objectives. U.S. Director of National Intelligence James Clapper, for example, noted in February 2016 that Russia would likely continue to "covertly use military and paramilitary forces in a neighboring state" and to "take information warfare to a new level" (Clapper, 2016, p. 18). The context of Russian activities in Georgia, Crimea, and eastern Ukraine has increased Western attention on new generation warfare, as discussed in the previous chapter. Similarly, Oscar Jonsson and Robert Seeley highlight Russian use of "full-spectrum conflict," including "subversion, physical and information provocation, economic threats, posturing with regular forces, the use of special forces, and the military intelligence coordinating paramilitary groups and political front[s]." They note that these tools "have been part of the Russian/Soviet lexicon of conflict for generations," and have been used recently in Crimea, Georgia, Moldova, and elsewhere (Jonsson and Seeley, 2015).

Russia's nonmilitary tactics can be understood as operating in a range of geographic concentric circles. Its interests, resources, and ability to deploy tools of foreign policy are reduced, though remain significant, in the outer circles.[21] In the first circle are countries directly within Russia's sphere of influence, including the Commonwealth of Independent States (CIS).[22] Analysts emphasize that Russia sees itself as a regional great power, and that this leads it to "influence the domestic politics of the surrounding states" (Saivetz, 2012, p. 402). Countries in Russia's immediate region, including Belarus, Kazakhstan, Krygyzstan, Tajikistan, and Uzbekistan, have continuing close economic ties with Russia, membership in Russian-controlled treaty organizations, and in some cases significant numbers of Russians or Russian speakers. With many former Soviet citizens and a parallel experience of post-Communism, they are also likely more receptive to Russian media. In the second circle are former Soviet or Warsaw pact countries, some with significant numbers of Russian speakers, who are no longer under Russia's influence but over which Russia retains interests (Estonia, Latvia, Georgia, Ukraine). This group is more exposed to the West and Western media, and generally far less receptive to Russian influence. A third circle includes the West and the rest of the world, especially countries that are perceived as threatening Russia or influencing its neighborhood.[23]

The typology in Figure 3.1 is one way to conceptualize Russia's influence over organizations in foreign countries. It suggests a five-category schema that categorizes entities ranging from those that are

---

21 See the "Regional Priorities" section of the 2013 Foreign Policy Concept, for example, which discusses "priority areas" as "the development of bilateral and multilateral cooperation with the CIS Member States," and then goes on to discuss its relations with Georgia, the Black Sea and Caspian regions, and then the European Union, NATO and the rest of the world (Ministry of Foreign Affairs of the Russian Federation, 2013).

22 The Commonwealth of Independent States (CIS) includes Armenia, Azerbaijan, Belarus, Kazakhstan, Krygyzstan, Moldova, Russia, Tajikistan, and Uzbekistan.

23 One metric of which sphere countries fall into is the price they pay for Russian gas (Abdelal, 2013, pp. 421–56).

Figure 3.1
Continuum of State Control of Nongovernmental Groups

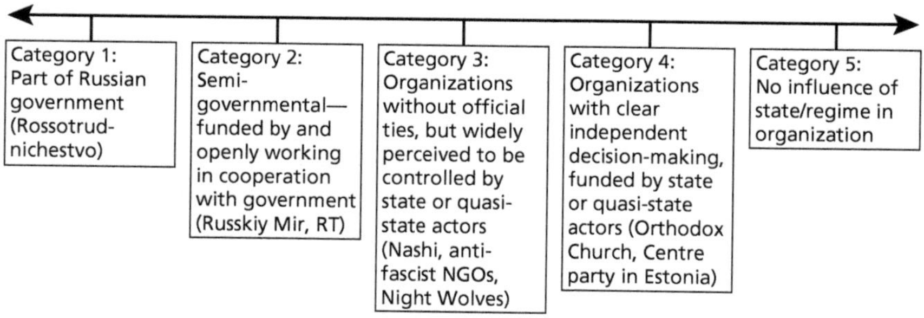

RAND *RR1772-3.1*

transparently part of the Russian state to those over which the Russian state exerts no influence. Given the extensive influence of the Russian regime in its neighborhood, it is difficult to definitively assign groups to these categories, but this typology offers a first cut at highlighting the different means and approaches of Russian influence.

Still, there is some risk of exaggerating Russian influence and capabilities. Despite claims that political technologists such as Surkov direct NGOs, for example, groups whose agendas align with Russia's goals could be legitimate and independent political organizations. Indeed, not everything that happens in Russia's favor is necessarily initiated by Russia, just as the United States is not responsible for everything that happens in its favor.

Diplomatic and Proxy

Russia uses formal (diplomatic) and informal (proxy) ties with a wide range of foreign organizations as key means of influence in foreign countries, especially in its near abroad, meaning Russia's perceived sphere of influence.

Russia has official relations with former Soviet republics through several formal institutions, including the CIS, the Collective Security and Treaty Organization (CSTO), and the Eurasian Economic Union (EEU). These institutions were created to manage the breakup or "divorce" of the Soviet Union, to compete with and counterbalance

Western institutions such as the EU and NATO, as well as to secure Russia's influence within its region. Analysts and observers have consistently questioned the ability of these organizations to achieve their declared goals of economic and political integration, and emphasized that Russia has extensive means of influence outside of these organizations. Nevertheless, these organizations retain a role in Russian foreign policy and a means of coordination and influence with its neighbors (Saivetz, 2015).

Russia's Compatriot Policy is a principal tool of its foreign policy in its near abroad. The policy was ostensibly developed as a means for Russia to protect and maintain a connection with its "compatriots," meaning Russians, former Soviet citizens, and their descendants living outside of Russia.[24] The Compatriot Policy was developed following the color revolutions in the early 2000s, in which governments friendly to Russia were forced from power in favor of Western-leaning governments. Marlene Laruelle observes that the regime believed that it was losing relative to the West, and took action accordingly to strengthen its influence within its near abroad through the justification of protecting compatriots.[25] Indeed, the Compatriot Policy has become more of a tool through which Russia can use Russian-speakers, former Soviet citizens, and affiliated groups to influence the politics of its neighbors, rather than a means to protect the interests of these individuals.[26]

---

24 The Compatriot Policy applies to four categories: Russian citizens living abroad, individuals who used to hold Soviet citizenship, individuals who migrated from the Soviet Union or Russian Federation, and descendants of the three previous categories except those who identify with their new home country.

25 Laruelle explains, "The Kremlin's perception that it was defeated in its neighborhood had serious repercussions and revived Moscow's will to invest into soft power and image-making. The Presidential Directorate for Interregional and Cultural Relations (Upravlenie Prezidenta Rossii po mezhregional'nym i kul'turnym sviaziam) was put in charge of conceptualizing Russia's new policy toward the Near Abroad. This directorate was led by another influential 'polit-technologist,' Modest Kolerov, who was dismissed after the 2007 Bronze Soldier scandal in Estonia" (Laruelle, 2015, pp. 9–10).

26 One report observes, "The so-called Russian compatriots, or Russians who live outside the territory of the Russian Federation, are becoming one of the most important Russian foreign policy tools to exert leverage on their host countries. . . . Russian foreign policy makers officially use the notion of protecting compatriots for purely decorative purposes" (Pelnens, 2009, p. 63).

Official government organizations, category 1 in figure 3.1, play a major role in the Compatriot Policy. The Ministry of Foreign Affairs has a Department for Cooperation with Compatriots. This department leads the Compatriot Policy in collaboration with Rossotrudnichestvo,[27] a separate agency under the Ministry of Foreign Affairs. Rossotrudnichestvo's full title is the "Federal Agency for the Commonwealth of Independent States, Compatriots Living Abroad, and International Cultural Cooperation," and it appears designed to mirror the U.S. Agency for International Development (USAID). As its full name implies, Rossotrudnichestvo's apparent mission is to expand Russia's influence abroad, with a focus on the CIS countries (Van Herpen, 2016, p. 36).

These agencies work in close coordination with category 2 organizations, which are formally nongovernmental but are state-funded and/or transparently work in close coordination with the state. The most prominent is *Russkiy Mir*, which was created by a presidential decree in 2007 with the goal of promoting the Russian language, culture, and education abroad (Sergunin and Karabeshkin, 2015, p. 352). Rather than an organization simply supporting cultural exchange, *Russkiy Mir* may instead be a key link between the Russian government and groups pursuing Russia's interests. Another example is the Institute for Democracy and Cooperation (IDC), with offices in Paris and New York, which acts as a Kremlin-funded think tank, focusing on "human rights based on traditional values" (Braw, 2015; Van Herpen, 2016, p. 34). Unlike the Western foreign aid organizations that the Russian organizations are claiming to emulate, the funding by Russian government actors remains confidential. In Latvia, for example, *Re:Baltica* identified 20 organizations that secretly received funding from *Russkiy Mir*, contrary to local laws that require NGOs to list the name of their funders (Spriņģe, Motuzaite, and Gailāne, 2012).

The regime often seeks influence through other organizations. Some of these are in category 3, formally independent organizations with no official ties but that are believed to be closely controlled proxies of the state. Perhaps the best example of a category 3 organization is the Night Wolves, a Russian biker gang with ties to Putin. Mark Galleotti

[27] See the Rossotrudnichestvo's homepage at http://rs.gov.ru/en.

writes that "the 5,000-strong Night Wolves are best thought of as auxiliaries of the state," noting that they have received significant amounts of money from the Russian government. He explains that Putin's support of the biker gangs reflects the Kremlin's broader strategy of co-opting potentially opposing groups: "they occupy the cultural niche that the real outlaws would otherwise colonize, and deny it to groups that would be rather less congenial for the Kremlin" (Galeotti, 2015). The Night Wolves have appeared in Crimea and elsewhere to voice their support for expanding Russian influence (one slogan of the gang is "wherever the Night Wolves are, that should be considered Russia") (Galeotti, 2015; Walker, 2015). There are also likely other organizations in this category in which the state's influence remains unknown or secret.

The Russian regime has a more complex relationship with category 4 organizations, which are fully independent but have an agenda that is somewhat in parallel with the regime. These organizations may receive some funding or have formal ties with the government or regime, but the regime's influence over their decisions is more limited. For example, political parties in the region—including Yanukovych's Party of Regions in Ukraine ("Party of Regions Hopes for Strengthening Collaboration with 'United Russia' Party," 2009), Harmony Centre in Latvia, and the Centre Party in Estonia—have ties with United Russia, the governing party in Russia (Milne, 2015; "How to Deal with Harmony," 2014). Similarly, a Russian bank has provided a large loan to the National Front in France, and Viktor Orban, the prime minister of Hungary, is reported to have links with Putin (Chazan, 2015; "The Viktor and Vladimir Show," 2015). Despite these ties, these parties are not necessarily under the thumb of the Russian government.

The Russian Orthodox Church, also arguably in category 4, serves as an example of the Russian culture that the soft power policy is intended to defend. The Russian Ministry of Foreign Affairs and the Church have a joint commission, Putin commonly speaks supportively of the Church, and some oligarchs close to Putin fund and have close ties with the Church. Priests from the Russian Orthodox Church have also been active with the separatists in Ukraine, while the separatists have attacked non-Orthodox believers within their territory. Though the Church is generally supportive of Putin, it has not marched in lockstep with the regime's policy, such as by not voicing strong support for the annexation of Crimea. The Church appears to be concerned especially with avoiding alienating believers in Russia's near abroad, and appearing too obviously as a direct proxy of the regime (Coyer, 2015).

A final group in category 4 are organizations led by wealthy individuals, often referred to as oligarchs. Though oligarchs may sometimes act as agents of the regime, they may sometimes seek to influence the regime or even act contrary to the regime's preferences. For example, Konstantin Malofeev, a Russian billionaire who supports many Orthodox Christian causes, is believed to have been instrumental in supporting Igor Strelkov and Aleksandr Borodai in creating a separatist movement in eastern Ukraine only initially (Weaver, 2014; Luhn, 2014). Indeed, the movement in eastern Ukraine did not appear explicitly to seek the goal of separatism until Strelkov and Borodai took over control (Weaver, 2014; Luhn, 2014). Later, however, Strelkov and Borodai left their positions in the leadership of the separatist movement, likely involuntarily, and were returned to Moscow, where Strelkov has become a more active critic of the regime (Kravchenko, 2015; Roth, 2015). While Russia has clearly offered support and at times encouragement to like-minded organizations, the regime faces principal-agent problems with groups that it supports.

Russia's influence over independent organizations is limited and specific to each context. For example, while Russia was able to nurture a separatist movement in eastern Ukraine, separatism did not spread to other cities in Ukraine, such as Odessa and Kharkiv, despite large numbers of individuals sympathetic with Russia (Pifer and Thoburn, 2014). Also, heavy-handed enforcement of Russian leadership over the separatists, including assassinations of separatist leaders, has been reported.[28] It should not be assumed that an organization is controlled by the regime simply because it has links with it; rather, assessing Russian influence over a given organization requires a case-specific assessment of the relationship and political context.

---

[28] Discussions with analysts, Ukrainian officials, and former OSCE observers, May 2015–January 2016.

Informational

The Russian government considers mass communication a vital battleground of international politics and has developed an extensive, well-funded media arsenal to facilitate its foreign and domestic objectives.[29] Two major agencies that work for audiences outside of Russia spearhead this work: Rossiya Segodnya and RT (see Figure 3.2). It is alleged that these agencies are separate entities, though there is overlapping leadership, including the same editor-in-chief, Margarita Simonyan.[30]

Rossiya Segodnya is part of the Russian government, a category 1 organization, intended to act as an official international information agency.[31] The agency's website states that its mission is to provide timely, balanced, and objective information of world events and share a variety of views on key events, including views not covered by other entities ([illegible], "[illegible] [illegible]"). The President of Russia appoints and fires the director-general of the agency, currently Dmitriy Kiselev. Kiselev is a popular Russian anchor known abroad for his radical anti-Western views. In his own words, he aims for the agency to "restore fair attitudes toward Russia, as an important global actor with good intentions" (Vesti.ru [News.ru], 2013).

Based in Russia, Rossiya Segodya combines radio broadcasting, news feeds, information platforms, multimedia international press centers, production and transmission of photo content and infographics, social media platforms, and content for mobile devices. While no official financial information is provided on the agency's website, other

---

[29] Accordingly, in its Foreign Policy Concept of 2013, Russia declared that it must "create instruments for influencing how it is perceived in the world," "develop its own effective means of information influence on public opinion abroad," and "counteract information threats to its sovereignty and security" (Russian Foreign Ministry, 2013).

[30] Most Russian government-funded platforms existed before the onset of the Ukraine crisis, but Russia's actions in Ukraine and the role attributed to Russian propaganda in the annexation of Crimea (see, for example, NATO Strategic Communication Center of Excellence, 2014) have brought to the forefront concerns about the potentially insidious influence of Russian media on international audiences.

[31] Rossiya Segodnya was established on December 9, 2013, by President Putin's decree on "Measures to increase effectiveness of state media" (see Ria Novosti, n.d.).

Figure 3.2
State-Funded Russian Media for Foreign Audiences

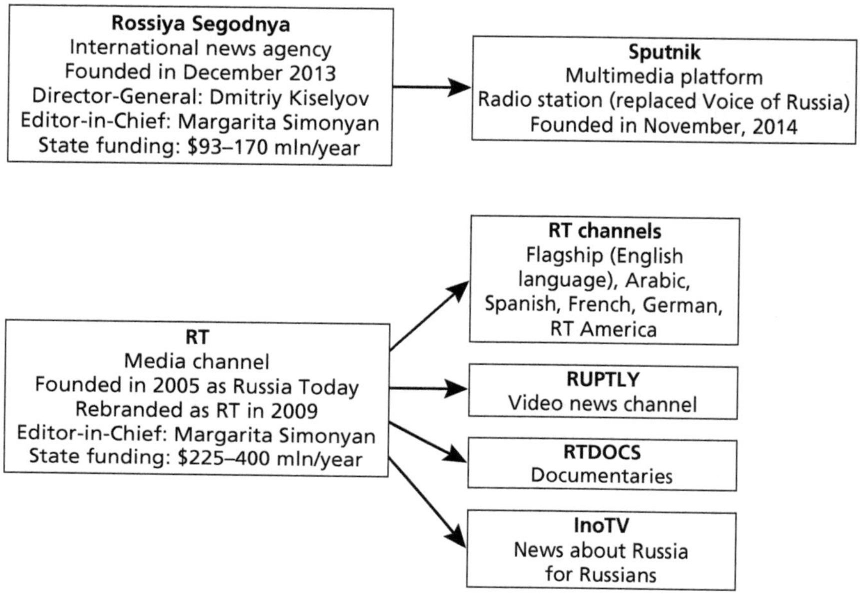

RAND *RR1772-3.2*

resources indicate that its annual funding reaches about $170 million (Tétrault-Farber, 2014).

Sputnik, an online and radio platform launched by Rossiya Segodnya in November 2014 to replace the Voice of Russia, broadcasts exclusively to foreign audiences. Sputnik claims to "point the way to a multipolar world that respects every country's national interests, culture, history and traditions" (Sputnik International, "About Us," n.d.). Sputnik includes websites, analogue and digital radio broadcasting, mobile applications, and social media pages; publishes materials in about 30 languages; and broadcasts news around the clock in English, Arabic, Spanish, and Chinese.

The Russian government created RT in December 2005 as a public relations instrument for promoting Russia's positive image abroad. Unlike Rossiya Segodnya, RT is a category 2 organization: It is registered as a nonprofit organization (TV-Novosti) and is funded through the Federal Agency for Press and Mass Communications. The organization was originally called *Russia Today*, and it focused mostly on cultural and news programming about Russia. Gaining little success abroad, in 2009 the network adopted a new name and refocused its coverage away from events inside Russia to delivering stories on global topics. Since then, RT's declared objective has been to "provide an alternative perspective on major global events and acquaint international audiences with the Russian viewpoint (RT.com, "About Us," n.d.).

RT has significantly expanded since its creation. Today, it has 22 bureaus in 19 countries and territories and employs over 1,000 media professionals (RT.com, "About Us," n.d.). It currently broadcasts through RT International (flagship channel in English), RT America (English), Rusiya Al-Yaum (Arabic), Actualidad RT (Spanish), and the documentary channel RTDoc. In 2013, RT added a Berlin-based, full-service video agency, Ruptly, which delivers news footage from global hotspots. RT also hosts InoTV, a channel that broadcasts in Russian the news coverage of Russia from other agencies and countries. The sharp decline of the ruble at the end of 2014 forced RT to postpone its opening of new channels in German and French (Shuster, 2015, pp. 44–51), although RT now operates news websites in German and French. All RT channels present around-the-clock news bulletins, documentaries, talk shows, and debates, as well as some programming about Russia's culture and sports. RT also boasts the popularity of its YouTube channel, where it is the number one TV news network, with over 2.5 billion views and nearly 3 million subscribers. RT sees its YouTube platform as "the core element of the network's emphasis on multi-media news delivery" and "high engagement with the audience" (RT.com, "About Us," n.d.). According to RT, its annual funding in 2015 was $225 million (RT.com, 2015). Other sources give much higher estimates (RT.com, "Facts vs Fiction," n.d.).

Besides these operations, Russian-language programming created for domestic Russian audiences is made available to the public across Russia's near abroad, where a good portion of the population understands Russian. Russian state-controlled platforms, such as Russia 24, NTV, Channel One, and RTR (Russia-1), as well as Russia's most popular social media platform, VKontakte, dominate the information space in the Commonwealth of Independent States and enjoy a nearly complete monopoly on Russian-language coverage in the Baltic States (NATO STRATCOM Center of Excellence, 2015).[32] Russian media in the near abroad has maintained its attractiveness to local audiences through diverse entertainment programming, interspersed with news.

The Russian government also appears to be sponsoring an intensive effort to influence social media through "troll farms." These groups regularly post information to websites in Russia and abroad to reflect the regime's point of view, cast doubt on Western narratives, or otherwise influence public opinion (Volchek and Sindelar, 2015).[33] While conclusive links between the "troll farms" and the Russian government have not been established, these groups are likely separate organizations funded by the state—i.e., category 3 organizations.

Russia's Media Messages and Target Audiences

Russia explains its media messaging as communication of its own legitimate viewpoint—it claims its messages counter the "Anglo-Saxon monopoly of information streams" ("Visit to 'Russia Today,'" 2013), as well as "restore fair attitudes toward Russia, as an important global actor with good intentions" and "provide a picture that is closer to reality" (Vesti.ru, 2013). To this end, RT and other news sources emphasize that they provide an alternate view to Western mainstream media and offer space to voices on the political and social margins, such as Julian Assange. RT has, at times, also produced high-quality and even award-nominated work (RT.com, 2012).

However, when it comes to international affairs, Russian media tends to feature mostly biased, agenda-driven programming, ridden with conspiracy theories, falsified facts, and selective, largely critical,

---

[32] Additionally, providers such as www.kartinausa.tv, which streams programs soon after they are available in Russia, or the Swedish Modern Times Group facilitate the channels' penetration to Lithuania.

[33] An article in the *Moscow Times*, for example, notes that Russian authorities first used manipulation of Internet content to influence domestic politics, and since "these tools were so effective at home . . . [first deputy chief of staff of the Kremlin Vyacheslav] Volodin's team decided to apply them to U.S. and European audiences as well" (Klishin, 2014).

reporting.[34] At the same time, it is mostly silent on corruption, suppression of free speech, or other negative stories from within Russia.[35] This is why many argue that Russian media abroad has been increasingly reminiscent of the Soviet-style propaganda, and is a method by which the Russian state competes with the West over Ukraine and other issues ("RT Is Very Upset," 2014).

Unlike its Soviet predecessors, however, Russian information efforts rarely aim to persuade or convince ideological opponents of an alternative truth. Instead, the apparent intent of Russian media seems to be to confuse narratives through the use of conspiracy theories and a barrage of alternative explanations, to undermine trust toward the established media and political institutions, and to instill a belief that the truth is unattainable, thereby making it more difficult to support or condemn actions of any particular actor (Snyder, 2014). A clear example of this was Russian media coverage of the tragedy involving flight MH17—an apparent attempt to drown the Western-proposed version of events in the cacophony of multiple unlikely possibilities ("The Kremlin's English-Language Media," 2014). The effects of obfuscation can be powerful even if the communications do not engender belief, and these effects can be compounded if not countered by concerted and authoritative refutation backed by evidence. When accused of biased reporting, Russian producers often point to the obsolescence of objectivity, assert that bias is inevitable in all media (especially Western),

---

34 The journalistic standards of Russian government-owned media have been questioned, especially during the Ukraine crisis. Britain's broadcasting regulator Ofcom issued a warning to RT in November, 2014 after concluding that it had "failed to preserve due impartiality" in four broadcasts about the conflict in eastern Ukraine. In another case, Ofcom sanctioned the RT program "Truthseeker" for misleading and biased programming on the conflicts in Syria and Ukraine, after which the program was canceled (Jackson, 2015; Bidder, 2015).

35 In a recent study, students at the Columbia University Graduate School of Journalism in New York spent two semesters monitoring RT. According to the project's research supervisor, Professor Ann Cooper, "on some subjects, especially what's happening in Ukraine, RT just completely toes the Kremlin line." Predictably, the study also noted the lack of criticism of Putin in any of the channel's programming. Further, the students were particularly surprised by the abundance of programming on the channel that would likely appeal to a U.S. or European audience "that doesn't trust government, is pretty cynical about government and is open to believing in certain conspiracy theories."

and claim that the reaction to Russian reporting is the West's reflection on its own flaws (TV Rain, Hard Day's Night, 2015).

Overall, Russian media mirrors the regime's generally fluid approach to ideology, which allows for the tailoring of messages in accordance with different needs and audiences. Both in Europe and in the United States, RT appeals mostly to the audiences on different ends of the political margins, similar only in their disillusionment with their governments and skepticism toward Western media. RT viewers simultaneously encompass the German right-wing fringe, the left-leaning British, the far right in France and Austria, America's far left, staunch conservatives, and libertarians (Shuster, 2015). As Pomerantsev and Weiss note, "European right-nationalists are seduced by the anti-EU message; members of the far left are brought in by tales of fighting U.S. hegemony; U.S. religious conservatives are convinced by the Kremlin's stance against homosexuality" (Pomerantsev and Weiss, 2014, p. 19). Such an approach to targeting and messaging indicates that Russian media puts great emphasis on understanding the political and societal landscape of each context where it broadcasts. The strategy appears to be not to create a new sentiment but to identify and tap into existing divides and grievances of the targeted audiences (Luxmoore, 2015), and then use them to promote Russia's views and facilitate its foreign policy objectives.[36]

Penetration and Impact of Russian Media

According to its website, RT has a global reach of over 644 million people in more than 100 countries and is available in more than 2.7 million hotel rooms.[37] These numbers, however, refer to reach, rather than actual viewership, which appears to be significantly smaller. According to the European News Market survey in 2013, RT was one of the top five pan-European news channels in terms of international reach within Europe (33 countries) (Kevin, Pellicanò, and Schneeberger,

---

[36] Media efforts complement Russia's other approaches to exerting influence in Europe, e.g., through the exports of natural gas and by cultivating relations with far-left and far-right parties in Europe, and with Russian investment.

[37] RT cites research by Nielsen, which RT has commissioned, but the results of which are not published anywhere.

2013). However, with respect to the actual viewership numbers, RT was lagging behind most other channels (last data available from 2012): Only about 0.5 percent of the European audience who watched news tuned in to get it from RT on a daily basis, about 3.5 percent watched it monthly, and 1.2 percent logged on to the RT website to seek news there. Further, according to the British Broadcasters Audience Research Board (BARB), in August of 2015 RT's weekly audience did not exceed 425,000 people, or 0.8 percent of total viewers in the UK (Broadcasters' Audience Research Board, 2016). These numbers remained largely the same since October 2014 (a month before RT launched a designated RT UK channel). These numbers are also about 100,000 fewer than in 2012, when BARB started measuring RT's UK audiences. Globally, RT's current viewership is only half the size of its main English-language competitor, Al-Jazeera English.[38]

RT's popularity in social media lags far behind that of the large broadcasters, like CNN or BBC. In 2014, RT's official Twitter account had 815,000 followers, which was barely one-tenth of BBC World News' 8.05 million followers, and orders of magnitude below CNN's 14.9 million followers. As of the date of writing, RT had 3,379,446 fans on Facebook, compared with BBC's 28,193,841 and CNN's 20,976,493.[39] The RT numbers are surprisingly low given the amount of effort the station has put into using the Internet to reach out to a younger, more tech-savvy audience. To increase its visibility and reach, RT tries to interconnect with other websites, particularly those on the extreme right and the extreme left (Armstrong, 2015).

While the strategy of attracting audiences on the margins of the global political spectrum may have increased overall viewership, it is

---

[38] In Germany, RT Deutsch currently works only as a web platform, which features one 30-minute daily news show, focused, in large part, on conspiracy theories and their perpetrators. Since the early days of its establishment in Germany, RT has acquired a reputation of being a biased propaganda channel, which made people reluctant to give interviews and otherwise partake in the channel's activities. RT has expressed intentions to create a full German TV channel, but so far there have been few signs that these efforts are moving forward. Sputnik does not quite have the audience in Germany, has failed to acquire a radio frequency, and currently broadcasts only in certain regions.

[39] Data assessed by the authors in March 2016.

unlikely to produce a coordinated response across different groups of viewers. The audience of channels like RT may pick and choose what issues it wants to hear about, particularly when getting information online (e.g., "How Americans Access Their News," 2014). An RT report may reach environmentalists whose opposition to fracking dovetails with the Kremlin's, but there is no reason to assume these groups are going to be receptive to the network's editorial line on Ukraine or other issues.

RT's emphasis on videos as a method of distributing information may have helped cement its status as the most popular news channel on YouTube, but it also means "it is broadcasting to a number of balkanized echo chambers" (Kennedy, 2014). As much as Russia may like to argue it is breaking the Western media monopoly, it is still a long way from being able to frame the debate as successfully as it is able to do at home.

For these and other reasons, Russia appears to have thus far failed to achieve its goal to boost its own image and undermine that of the United States. A 2015 Pew Research Center survey has shown that "a median of only 30% see Russia favorably in the nations outside of Russia. Its image trails that of the United States in nearly every region of the world. At the same time, a median of only 24% in the countries surveyed have confidence in Putin to do the right thing in world affairs, and there is far less faith in the Russian leader than there is in U.S. President Barack Obama" (Stokes, 2015). Opinions of Russia are more unfavorable than favorable in 26 nations, with anti-Russian sentiment particularly strong in Israel (74 percent), Japan (73 percent), Germany (70 percent) and France (70 percent) (Stokes, 2015).

However, Russian information efforts in its near abroad appear to have been more successful and farther reaching. A great majority of Russian speakers in former Soviet republics consume Russian-produced or Russian-sponsored programming (BBC, 2015), sometimes because their countries offer few alternatives in the Russian language. Through its developed and long-standing media networks in the former republics, Russia is able to maintain strong connections with its "compatriots" and shape their daily narratives both on Russia and on the countries where they live. This established rapport and connectedness then becomes particularly important in times of crisis, as Russia is able to use its channels to transmit the Kremlin's line. During the 2014 Crimean crisis, for example, the Russian media used its extensive presence on the peninsula to paint its own image of the events in Kyiv to aggravate the Russian Crimeans' existing sense of estrangement from and grievances against Ukraine, and their nostalgia for Soviet unity with Russia ("Opros" [Survey], 2013). Similarly, in the Estonian Bronze Night crisis discussed below, Russian media had a variety of established and popular channels to promote its own take on the events.

To sum up, Russia has invested significant resources and efforts in developing a diverse and sophisticated set of information channels aimed at promoting Russia's goals abroad: improving its image, framing discussions in a way that serves Russian interests, undermining the legitimacy of Western and Western-leaning governments and institutions, etc. The success of Russian international information efforts has varied from context to context, with greater impact achieved in Russia's near abroad than in the West. However, given an increased emphasis within Russia on "hybrid threats" and "informational warfare" (Voennaja doktrina Rossijskoj Federacii ☒☒☒☒☒☒☒ ☒☒☒☒☒☒☒☒ ☒☒☒☒☒☒☒☒☒☒ ☒☒☒☒☒☒☒☒☒ [Military Doctrine of Russian Federation]), Russia will most likely continue to hone and expand its information tools of influence both close to and far beyond its borders.

Cyber

Russia appears to be increasingly using cyber attacks as part of its political-military campaigns. Cyber warfare can be understood as one component of Russia's understanding of information warfare. As the Estonia case study below details, cyber attacks against the Estonian government escalated following the Bronze Night incident in 2007. In Georgia in 2008, official websites were defaced; denial-of-service attacks were launched against government, financial, and news websites; and Georgian politicians' e-mail addresses were spammed (Tikk, Kaska, and Vihul, 2010, pp. 71–74). In Crimea in 2014, there were "low-tech cyberassaults on over a hundred government and industrial organizations in Poland and Ukraine, as well as attacks on the European Parliament and the European Commission" (Matthews 2015). Russia also seized civilian telecommunications infrastructure to control the information entering and leaving Crimea (Gady, 2014; interview with think tank analyst, 2016). In January 2016 in eastern Ukraine, there were accounts of a cyber attack against the regional power authorities, leading to a major power outage (Goodin, 2016). In July 2016, several cyber security firms and researchers claimed that Russia was likely behind cyber attacks on the Democratic National Committee, based on identifying malware previously attributed to Russian groups and evidence of the use of the Russian language in the software used in the attacks (Alperovitch, 2016; Glaser, 2016).

Russian cyber warfare capabilities draw from organizations both within and outside of the government. The Federal Security Service (FSB), Russia's main domestic security agency, incorporated significant electronic intelligence resources from the now-disbanded Federal Agency of Government Communication and Information (FAPSI). The FSB appears to have the means for monitoring the use of the Internet within Russia and abroad, and is believed to have other cyber warfare instruments as well (Heickerö, 2010, pp. 27–31). While there has not been fully conclusive confirmation of attribution of the groups responsible for the DNC attacks, two groups, which did not appear to coordinate their efforts, have been identified: APT 28, which appears to be associated with the GRU, and APT 29, which appears to be linked to another Russian intelligence organization. These groups are also blamed for a range of other attacks against government and civilian targets in the United States and around the world (Alperovitch 2016).

Though there is growing evidence for suspecting Russian involvement in the DNC and other attacks, in general attributing responsibility to government actors is difficult. There appear to be many civilian hackers who are employed by or controlled by the state (category 3) or who are independent but whose interests sometimes align with the state (category 4). Cyber expert Jeffrey Carr notes that "hundreds of 'black-hat' Russian hackers" work either "at the order of Swiss bankers or Ukrainian oligarchs." Carr also notes that many of these hackers are coopted by the FSB: "Russian hackers who are caught are given the choice to work for the FSB or to go to jail. The FSB also has some on contract hire" (Herzog, 2011, pp. 53–54).

Military

Russia has launched a major effort to modernize its military, estimated at $700 billion up to 2020 when first announced in 2011 (Trenin, 2014). One of the priorities of this reform appears to be developing more professional forces focused on low- and mid-intensity conflict, whose mission includes tasks under the rubric of political warfare (Bartles and McDermott, 2014). These forces appear to have been used in Crimea, eastern Ukraine, and other conflicts. Russian military activities related to political warfare also include the use of overflights, exercises, and nuclear rhetoric.

A first element of Russia's military capacity for political warfare is its specialized light infantry and rapid response forces. These include forces from a number of different organizations. One such group of forces is the airborne Vozdushno-Desantnye Voiska (VDV), composed of approximately 35,000 personnel. The VDV is an elite, mechanized force, equipped with infantry fighting vehicles and tanks, and includes parachute and air assault units. The VDV, as Bartles and McDermott note, "also act as a reliable enforcer of politically sensitive operations" (Bartles and McDermott, 2014, p. 52), dating back to Hungary in 1956. In Crimea, the Thirty-First Air Assault Brigade, which has special training as a peacekeeping unit, was reportedly flown to Sevastopol. The VDV has a relatively higher portion of contract soldiers and has benefitted from recent equipment upgrades (Bartles and McDermott, 2014, pp. 51–52, 57).

Russia also has a number of Spetsnaz (an abbreviation of *sily spetialnogo naznachenya*, or "special task forces") units under various military and law enforcement organizations.[40] Russian Spetsnaz forces include those with specialized training in a particular area (e.g., engineers or communications specialists) in addition to forces that conduct clandestine or sensitive missions.

---

40 For a complete list of Spetsnaz forces, see Nikolsky, 2014.

Spetsnaz appears to have a likely significant role in future political warfare. Igor Sutyagin, an analyst at RUSI, noted, "They have a range of uses. They can fight but they are also trained for intelligence work. To establish insurgencies. To control them. To smuggle arms. To wage guerrilla wars" (Jones, 2014). Similarly, Mark Galeotti identifies future roles for Spetsnaz, including acting as the " 'tip of the spear' . . . to clear that way for less nimble forces," counter-insurgency, and acting as a "political operator" in the "same category as the intelligence services, propaganda machine, and other agents of what the Soviets called 'active measures,' or political warfare" (Galeotti, 2014b, pp. 9–11).

The main military Spetsnaz forces are the seven Spetsnaz brigades, which are tasked primarily with providing special reconnaissance and intelligence. The training and doctrine of these units is under the Russian military intelligence agency, the GRU, though they are operationally subordinated to Russia's Military Districts and Fleets.[41] Unlike Western SOF, *spetsnaz*-GRU appear intended mainly to provide reconnaissance in support of Russian conventional forces,[42] as they did in Chechnya and Georgia (Bartles and McDermott, 2014, p. 50, 52–53). In 2013, a Spetsnaz Special Operations Command (Komanda Spetsial'nikh Operatsiy, or KSO) was created both as a separate command within the General Staff and as a proponent for all special operations forces (Bartles and McDermott, 2014, pp. 47–52). The apparent intent was to emulate the U.S. Special Operations Command by creating a more capable force directly under the operational control of the General Staff. Accounts of the KSO differ as to its size, ranging from 500 to 1,500 (Galeotti, 2014b; Bukkvoll, 2015, pp. 611–12).

---

[41] The Spetsnaz under the GRU were extensively reorganized as part of the Russian military reform following the war in Georgia in 2008. This reorganization included reducing the number of Spetsnaz brigades from nine to seven, reducing overall size of the *spetsnaz*-GRU, and the operational resubordination of the *spetsnaz*-GRU to the military districts. Tor Bukkvoll describes the reorganization as a significant "decline" for the *spetsnaz*-GRU (see Bukkvoll, 2015, pp. 611–14).

[42] One article notes, "In short, Russian Spetsnaz brigades are doctrinally and logistically bound to the armed forces and are not in-tended for the kind of independent operations that Western special operation forces conduct" (Bartles and McDermott, 2014).

There are detailed accounts of the role of Spetsnaz in the Russian operation in Crimea, which demonstrate some of these roles. Barles and McDermott write that Spetsnaz "almost certainly arrived well in advance to secure the local support necessary for the guise of 'polite people.'"[43] Mark Galeotti reports that Spetsnaz forces conducted "covert negotiations with members of the local ethnic Russian elite, who arranged for masked and armed 'self-defence militias' (whose numbers included both local police and criminals) to begin appearing on the streets." Subsequently, Spetsnaz "operators from the KSO and 45th opSpN were moved into the barracks of the 810th Independent Naval Infantry Brigade at Sevastopol. On February 27, these forces seized the Crimean parliament building and began blockading Ukrainian bases in Crimea" (Galeotti, 2014b, pp. 6–8). Accounts of Crimea emphasize the professionalism of the Spetsnaz, including their ability to seize critical sites while pretending to be irregulars and avoiding harm to the local population (Bartles and McDermott, 2014, pp. 57–59).

Reports of Spetsnaz activities in eastern Ukraine are less well demonstrated. There is some evidence of the presence of GRU forces, though little evidence about their activities. Sutyagin and Galeotti offer accounts of the presence of elements from the 2nd, 10th, 22nd, and 24th Spetsnaz Brigades and from the 45th Guards Spetsnaz Regiment of the VDV (Sutyagin, "Russian Forces in Ukraine," 2015; Galeotti, 2014b, p. 8; Bartles and McDermott, 2014, p. 57). Galeotti also reports an account of "a GRU unit that was possibly an ad hoc Spetsnaz force" being used to influence the separatist leadership (Leviev, 2015; Standish, 2015). Other analysts observe that the GRU has gained the increasing trust and respect of the regime because of its effectiveness in the war in eastern Ukraine (discussions with think tank analysts, 2016).

Bartles and McDermott highlight the role of Spetsnaz as part of a larger effort to create a 78,000-strong Russian Rapid Reaction Force comprising elements of the VDV, Spetsnaz, air, and sea forces. They

43 They further note, "One source we interviewed believed that select elements of five of Russia's seven GRU Spetsnaz brigades (the Second, Third, Tenth, Sixteenth, and Twenty-Second) and one GRU Spetsnaz regiment (the Twenty-Fifth) participated in the campaign" (Bartles and McDermott, 2014, p. 57).

emphasize that these forces have disproportionately benefitted from an increase in resources from the Russian military reform effort. They write that "the funding priority is smaller lighter, and more mobile forces capable of handling the threats Russia is now more likely to face: small-scale regional conflicts, terrorism, proliferation, and insurgency. . . . In short, the majority of resources is being funneled to a minority of the land fighters" (Bartles and McDermott, 2014, pp. 48–49).

A second military element of Russia's political warfare activity has been its extensive air and naval activity in and around NATO territory. NATO fighter aircraft in the Baltics moved to intercept Russian aircraft some 130 times in 2014. In one example, Sweden deployed naval forces after reports of a Russian submarine in their territory (Kramer, 2015; Frear, Kulesa, and Kearns, 2014). Steve Covington, a political advisor to the Supreme Allied Commander Europe (SACEUR), connects these activities to a larger strategy of breakout from Western encirclement: "Russia's strategic nuclear aviation flights and maritime deployments can be seen as the modern day versions of tactical breakout actions from World War II, designed to demonstrate the Kremlin's resolve to not accept encirclement."[44]

A third critical military component of Russia's political warfare is exercises, both scheduled and unscheduled. For example, in Zapad in 2013, a large number of troops (declared at 13,000 but estimated at as many as 70,000) exercised a response to attacks by "terrorist" forces from imaginary countries to Russia's west. In addition to conventional military forces, Zapad 2013 also involved participation of special forces from the Ministry of Internal Affairs and the FSB, implying a growing desire over previous exercises to develop coordination across the Russian government. Previous exercises included a simulated nuclear strike against Warsaw and the rapid deployment of 160,000 troops (Zdanavičius and Czekaj, 2015).

---

[44] Covington also connects these military deployments to other political, economic, and information elements as part of a strategy to "break with the norms, rules, practices of the post–Cold War period and destabilize the current security system" (Covington, 2015, pp. 9–10).

These exercises fulfill the conventional military priorities of training and the development of new capabilities. Exercises may be especially important to Russia given that a large portion of Russia's military is composed of one-year conscripts who are not always well integrated into their units. But these exercises also likely serve a clear political purpose. Ieva Berzina claims that Zapad 2013's main goal was strategic deterrence. Russia simultaneously emphasized the defensive character of the exercise while demonstrating Russia's extensive military capability. Russian leaders may have hoped that this would "convince potential aggressors that putting any form of pressure on Russia or its allies is futile" (Berzina, 2015). Nevertheless, Baltic officials saw the main purpose of Zapad 2013 as intimidation, pointing to the potential of these exercises to gain leverage over Russia's neighbors (Berzina, 2015).

Large-scale exercises and other deployments may also act as a prelude and cover for surprise military activities, as in the case of Crimea in 2014. On February 26 and 27, President Putin announced training drills and naval exercises involving around 50,000 forces that served as cover for airborne forces, naval infantry, and *spetsnaz*-GRU to move into Crimea (Hurt, 2014). Russia also deployed varying numbers of forces (20,000–40,000) across the border from Ukraine as the separatist campaign developed in mid-2014 (Jones and Olearchyk, 2014). These forces may have been intended to act as a strategic deterrent to dissuade Ukraine from deploying its military to intervene in the developing conflict. If so, this effort failed. While the Ukrainian military was incapable of deploying substantial combat forces to its eastern provinces early in the summer, by August it had mobilized and began to rapidly recapture territory from the separatists.[45] Russia was forced to deploy its own forces to prevent the defeat of the separatists in late August 2014, and as of May 2015, Ukrainian officials claimed some 55,000 Russian forces remained in the area, with 8,760 on the territory of Ukraine (discussions with Ukrainian officials, 2015; Sutyagin, 2015).

45 One Ukrainian officer observed that the troops were unprepared to even drive over a field of newly planted wheat in the spring of 2014, much less to actually fight the separatists. Only the newly created National Guard was able to mobilize significant troops. Interviews with Ukrainian officers, Kyiv, April and May 2015. (See also Finley, 2014; Faiola, 2014.)

Intelligence

Russia's intelligence community is a critical component of its political warfare efforts. Intelligence agencies are believed to be responsible for a range of covert activities that can be integral to Russian political objectives.

Following the dissolution of the Soviet Union in 1991, Russian intelligence agencies went through a significant period of reform and reorganization. The KGB dissolved into component parts, which then went through a number of changes and reforms. Many former KGB officials took advantage of the Communist Party's resources and rose to prominent positions in politics or business (Chrissikos, 2014). For example, Vladimir Putin was a KGB officer and maintained strong links with a number of other former KGB officials to advance his career following the end of the Soviet Union (Dawisha, 2015).

Four major organizations evolved from the Soviet era institutions. First, the FSB is Russia's main domestic security agency, with approximately 286,000 personnel (down from 386,000 in 2010). Though the FSB operates primarily within Russia and the CIS (Dawisha, 2015), where it has extensive offices and networks, it also has some limited responsibility in the rest of the world, especially for politically sensitive operations. Analysts emphasize, for example, that the FSB is more likely to be responsible than are the other intelligence agencies for activities such as bribery or subversion (discussions with analysts, 2015).

Second, the Foreign Intelligence Service (SVR) is the successor to the KGB's foreign intelligence organization. With around 15,000 personnel, the SVR focuses on the collection of information, with activities in Western Europe and elsewhere around the world (Jane's Sentinel Security Assessment, 2015).

Third, the Main Intelligence Directorate of the General Staff (GRU) is Russia's military intelligence organization. The GRU is believed to have similar size and resources as the SVR. The GRU is divided into the *agentura*-GRU, which includes the signals intelligence and personnel stationed at embassies, and the *spetsnaz*-GRU, discussed above in the military section (Bukkvoll, 2015, pp. 617–18; Jane's Sentinel Security Assessment, 2015).

Fourth, the Federal Protection Service (FSO) is responsible for top-level communications (Heickerö, 2010, p. 28). There remains significant competition over resources, responsibilities, and influence among these agencies (see Galeotti, 2016).

Intelligence organizations are often used for covert action that may have political purposes. The activities of Russian intelligence personnel in eastern Ukraine offer an example of how covert action supports Russia's sponsorship and control over the separatist movement. Strelkov, for example, claimed to be a reserve colonel in the FSB until March 2014, and FSB "proconsols" are likely active in the political organization of the separatists. Analysts also note significant numbers of GRU officers in separatist controlled areas of eastern Ukraine, and speculate that the GRU is likely responsible for the murder of separatist officials, thereby maintaining Russia's control over the movement (BBC News, 2014; discussions with analysts and former OSCE officials, 2015–2016). The recent British investigation revealing that FSB agents were responsible for poisoning Alexander Litvinenko in London in 2006 also highlights the potential involvement of Russian intelligence in violent covert activities (Litvinenko, 2016).

Intelligence agencies and functions play critical roles in several of the political warfare activities described above. The FSB appears to have a lead role in cyber warfare (discussion with think tank analyst, 2015; Heickero, 2010, p. 4). Military activities described above, including special reconnaissance, peace support, and military assistance, also depend on the support of the intelligence services. Indeed, one can think of the *spetsnaz*-GRU as performing both intelligence and military functions. Furthermore, reports have been made of intelligence services working closely with proxies, influencing propaganda, and conducting other tactics of political warfare. In Crimea before the war in Ukraine, for example, Taras Kuzio reports "covert support to separatist, anti-NATO and anti-American groups in the Crimea and Odessa," and notes reports by the Ukrainian National Security and Defence Council about Russian intelligence efforts "to discredit the Ukrainian leadership with the aim of reintegrating Ukraine into Russia's sphere of influence" (Kuzio, 2010, p. 30). Close links between Russian-funded organizations and Russian intelligence have also been reported. For example, Leonid Reshetnikov became the head of the Russian Institute for Strategic Studies after a three-decade career in Russian intelligence, and the Estonian Internal Security Service (Kaitsepolitsei, hereafter KAPO) reports that former Russian intelligence officials are active in the *Russkiy Mir* Foundation in Estonia (Braw, 2015, p. 31; Conoley and Gerber, 2011).

Economic

Russia has extensive economic ties in Europe and around the world, which it has repeatedly attempted to use for political influence, with mixed results.

Russia possesses a wide range of economic ties with its neighbors. For example, Russia is the world's second largest exporter of oil and gas after Saudi Arabia, the EU's third largest trade partner, and the dominant economic actor in the former Soviet Union (European Commission, 2016; Bordachev, 2015). Many Russian companies have substantial investments in Europe and around the world.[46] The nature and strength of Russia's economic ties, and the dependence of trade partners on Russia, vary extensively across industries and across the partners. Russia tends to have the closest economic ties with the former Soviet and Warsaw Pact countries, in part because of shared infrastructure from the Soviet period. In countries such as Bulgaria, Russian companies build on Soviet-era ties to gain control over major economic assets and expand Russian economic interests, especially through the development of energy production, processing, and distribution (Bechev, n.d., p. 5). In the early 1990s, the Baltic states also had close links with Russia and Russian organizations, including through energy imports, organized criminal networks, and shared transportation infrastructure. Indeed, some Baltic analysts highlight Russia's economic influence as a security threat (Spruds, 2012, pp. 40–49). The Baltic states, however, have taken significant measures to reduce their dependence and connections with Russia, including targeting Russian organized crime,

---

[46] Ongoing research by several organizations seeks to track and highlight the activity of Russian companies around the world, given the potential for economic and political influence (see, e.g., "IntelTrak," n.d).

diversifying energy supplies, building rail networks to European standards, and encouraging investment by Nordic banks (Hanley-Girsch, 2009; Schaefer, 2011; interviews with Baltic officials, July 2015). In Western Europe, there are some key sectors and businesses who depend on trade with Russia. For example, some agricultural businesses have significant trade with Russia, and were hit hard by Russia's 2014 embargo. In Norway, for example, sales to Russia amounted to 11 percent of overall agricultural exports. Similarly, German automakers have significant sales in Russia (Geiger, 2014; Rapoza, 2015). However, trade with Russia made up only 6 percent of overall trade volumes within the EU in 2015 (European Commission, 2016). Russia's economic links with European countries, and hence its potential tools of influence, thus vary extensively based on factors such as the presence of preexisting infrastructure (e.g., gas pipelines or rail networks), the pursuit of diversification, and whether there is parallel political influence present to support economic levers (see also Larrabee et al., 2017).

Variation across these factors leads to important differences in the vulnerability of countries to a cutoff of Russian gas, a scenario that is often cited as a major potential Russian means of economic warfare, especially as crude oil and refined products are far more easily fungible across different suppliers (Larrabee et al., 2017). While many European countries receive a significant portion of their energy supply from Russia, the possibility of economic harm from the cutoff depends on the availability of alternative supplies, use of alternative energy resources, and flexibility to decrease consumption. In October 2014, the European Commission released a study of how a six-month cutoff of natural gas from Russia would impact the security of supply in European countries. According to the study, as shown in Figure 3.3, Finland, Estonia, Bulgaria, and the western Balkan countries were highly vulnerable and would experience shortfalls of 80–100 percent of natural gas supply, while the supply of natural gas would largely be uninterrupted in Western Europe. The study also found that sharing of natural gas resources across European countries would significantly reduce the impact in both southern and northern European countries (European Commission, 2014). Still, Russia remains a major supplier (in particular, Russia was the source of 28 percent of Europe's natural gas in 2015 (Eurogas,

Figure 3.3
2014 Vulnerabilities to a Short-Term Shutoff of Russian Gas

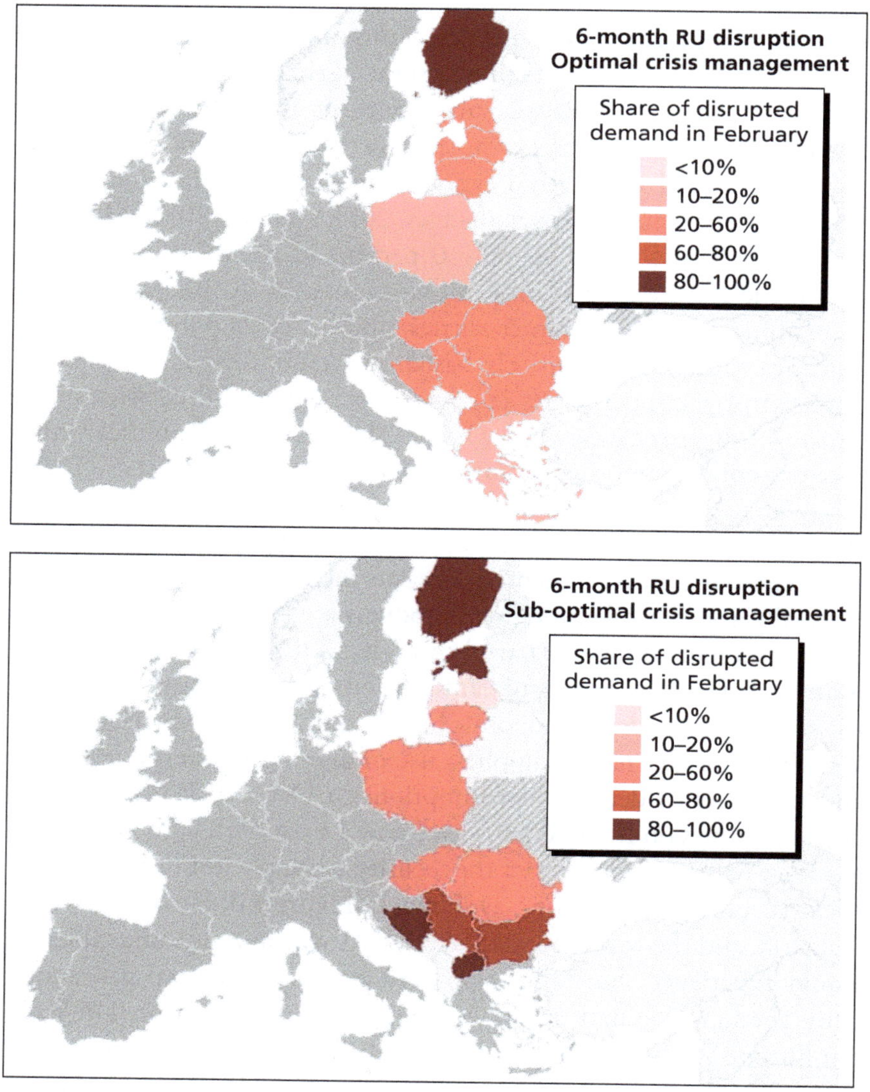

SOURCE: European Commission 2014, pp. 6.
RAND *RR1772-3.3*

2015)), and the long-term reduction in supply of gas from Russia would likely have significant repercussions for European energy. The EU has sought to reduce the vulnerability of European countries to a shutoff of Russia gas through rules encouraging diversification of imports, construction of interconnectors and new pipelines, and increases in production from non-Russian producers (Simon, 2015; European Commission, 2015). Even Ukraine, which was highly vulnerable to a cutoff of Russian gas, has significantly reduced its vulnerability since 2014.[47] Still, while the vulnerability of Europe to a cutoff of energy supplies from Russia is declining, Europe continues to rely heavily on energy imports and other forms of trade with Russia.

Russia's manipulation of these economic resources is facilitated, and to some degree complicated, by the close connections between Russian economic actors and the Russian state. For example, wealthy oligarchs depend on protection from and ties with the state, and may at times act on behalf of the government, while senior officials in the Russian government may take advantage of their positions to exercise their own businesses interests (Hill and Gaddy, 2013, p. 226; Pavlovsky, 2016). Jeffrey Mankoff (2009, p. 79) notes that the existence of what some refer to as "Kremlin, Inc."—"the nexus between wealth and power"—has significant foreign policy implications, especially creating an incentive for the Russian government to act in the interest of companies with close ties with the state. Further, some partially or fully state-owned companies—including notably the majority state-owned natural gas exporter, Gazprom—may at times act as a direct proxy of the Russian state (Abdelal, 2013).

The links between Russian companies and the state has at times facilitated Russia's political goals. For example, following the Orange Revolution in Ukraine in 2004, Gazprom significantly raised the prices for natural gas to Ukraine (which had previously purchased gas from Gazprom at significantly lower prices than Western Europe) at the same time as Ukraine experienced declining relations with Russia.

---

47 In 2015, Ukrainian gas imports from Russia went from 14.5 billion cubic meters (bcm) to 6.1, with increased imports from Europe and a decrease in consumption (Varfolomeyev, 2016).

Following the repeated failure of Ukraine's natural gas provider to pay its bills, Gazprom has several times cut gas supplies to Ukraine, in 2006, 2009, and 2013. Since much of the gas transiting Ukraine is meant for European customers, these interruptions in supplies at times impacted the supply of gas to other European countries. Gazprom's activities may have to some degree reflected the goal of increasing revenues from Ukraine, but they also reflected Russia's political interests in not subsidizing Ukraine as it moved closer to the West. Indeed, Russia initially appears to have demanded that Ukraine extend the lease on the naval base in Sevastopol and hold a referendum prior to any NATO accession in return for less expensive gas. Further, at the same time as Gazprom was increasing the price paid by Ukraine, Gazprom continued to provide significantly cheaper gas to other former Soviet republics (Mankoff, 2009, pp. 230–33; Abdelal, 2013; Blackwill and Harris, 2016).

In other cases, Russian companies' presence abroad remains substantial but the political impact is less certain. For example, in Bulgaria, the Russian oil company Lukoil owns a major refinery and controls a significant portion of the country's liquid fuel market. Several Russian banks have also made significant investments in the country (Bechev, n.d., pp. 5, 11–19). Analysts in Bulgaria claim that these and other Russian economic levers lead to political influence through the financing of political parties and other organizations, but note that Russia's activities in Bulgaria appear limited to continuing to extract rents and enlarging its economic presence (discussions with Bulgarian analysts, June 20–21, 2016).

Finally, in some cases the Russian state more directly sought to use its economic influence to achieve political outcomes through sanctions and other measures. Russia, for example, issued counter sanctions against European imports of agricultural products in 2014 following Europe's Crimea-related sanctions on Russia. In 2013 and 2014 in Ukraine, Putin reportedly threatened Viktor Yanukovitch, the President of Ukraine, that if he signed an Association Agreement with Europe then Russia would cut off trade with Ukraine, with catastrophic consequences for Ukraine's economy ("Summit of Failure," 2013; Schenkkan, 2014). Similarly, when Moldova pursued and eventually signed an Association Agreement with Europe in 2013 and 2014, Russia introduced sanctions on a number of Moldovan products, including wine, a major export (Herszenhorn, 2013; Calus, 2014). These examples demonstrate the willingness of Russia to adopt economic measures to achieve its goals, though not the effectiveness of these measures. In the case of the agricultural countersanctions against Europe in 2014, the sanctions against Russia remain intact, while in the cases of both Ukraine and Moldova, Russia failed to achieve its objectives of discouraging further integration with the EU.

Russia has substantial economic ties with Europe and its neighbors, and has periodically sought to use these economic ties to achieve its political goals. However, there are significant limits to Russia's economic influence, given its economic dependence on energy exports, and the alternatives available to many of its partners. Hence, though Russia can and likely will use its economic connections as a lever in the future, the effectiveness of these measures will vary depending on the circumstances.

Coordination/Leadership

Russia's ability to execute political warfare depends on coordination across the whole of its government. The literature on new generation warfare emphasizes the need to use a range of elements of statecraft and hence improve coordination across different agencies.[48]

Analysts, however, are generally skeptical about the ability of Russian organizations to easily coordinate their efforts. Russian agencies are often described as separate bureaucratic kingdoms that often compete for resources and turf within the government. These agencies can be brought together at times of national crisis, but on a day-to-day basis may have little incentive to cooperate effectively (discussions with

48 For example, one report on information warfare explains, "It is repeatedly stressed in official documents, as well as in military theory, the resources of many different government agencies need to come together to wage successful information war," including "the Federal Security Service (FSB), the Foreign Intelligence Service (SVR), the Armed Forces, the Military Intelligence Service (GRU), the IT and mass media supervision service Roskomnadzor, the Federal Protection Service (FSO), and the Ministry for Foreign Affairs" (Franke, 2015, p. 51).

U.S. think tank analysts, 2015). Georgiy Filimonov, the director of the Russian Institute of Strategic Studies and Forecasts, for example, observes that Russia lacks an established mechanism of cross-institutional coordination of its soft power efforts. He writes that Russia's public diplomacy is manually directed by the country's leadership and administered through the president's press services and other branches of his administration, as well as through the Foreign Ministry and Rossotrudnichestvo (Yu, 2013). The head of Rossotrudnichestvo, Konstantin Kosachev, admitted that Russia's public diplomacy suffers because of the lack of coordination and insufficient funding ("Rossotrudnichestvo razrabotalo doktrinu 'myagkoy sily'" [Rossotrudnichestvo Developed the "Soft power" Doctrine], 2014).

Signs of improving coordination, however, can be seen, especially in the operation in Crimea.[49] The National Center for the Management of Defense of the Russian Federation was recently established under the General Staff. The center appears intended to facilitate the coordination of the military with a range of nonmilitary organizations, including the Ministry of Internal Affairs, Ministry of Emergency Situations, and others. Roger McDermott cites statements from Russian officials explaining that "the driving force in creating the new center is to achieve integration of information flows into a single channel—something unique in Russian defense and security experience" (McDermott, 2014). Similarly, Rossotrudnichestvo proposed a draft strategy document intended to improve coordination among state organizations engaged in strategic communications ("O sisteme koordinacii [On a System of Coordination], 2014). Some analysts highlight the role of Russia's National Security Council, under the presidential administration, while others note the likely leading role of the FSB,

[49] Samuel Charap writes, "Most significantly, Moscow coordinated the arms of national power effectively in order to achieve its objectives. In previous military operations in the post-Soviet period, Russia was not able to do so. The operation in Ukraine included the successful use of subversion, cyber, proxies, conventional military interventions and military exercises to deter and coerce, all conducted under the cover of the nuclear umbrella, which Russian officials regularly brought to the world's attention" (Charap, 2015, p. 53).

especially for coordinating and leading Russia's control of information (discussions with U.S. think tank analysts, 2015).

There appear to be differing views about the precise mechanisms by which the regime executes its preferred policy decisions. Russia has a strong "power vertical," meaning the hierarchy under the president ensures that policy follows his preferences, but some have questioned the efficiency and consistency of the Russian hierarchy. For example, in September 2014, the FSB arrested KAPO officer Eston Kohver, claiming that he was on the Russian side of the border engaging in espionage from Estonia. Estonian sources instead claim that he had been investigating organized crime involving FSB agents, and some analysts suggest that the FSB may have acted without the full authorization of the Kremlin in this case (discussions with think tank analysts, 2016).

On one hand, the centralization of Putin's control enables rapid and effective decisionmaking by the Russian state, which is clearly beneficial for effective coordination of political warfare, among other policies. On the other hand, centralization under Putin has not been entirely effective, as the presidential administration is able to coordinate only a limited number of activities (discussions with thank tank analysts, 2016). In this sense, the more that Putin directly controls, the more difficult it is to ensure that a particular government agency is following his instructions. Hence, although Russia does have a wide range of available tactics for political warfare and is pursuing increasing efforts to improve the links among government organizations, effective coordination of political warfare will remain a significant challenge.

Russia's Political Warfare in Estonia

Estonia, though small (with a population of 1.3 million), offers a useful demonstration of the capabilities and limits of Russian political warfare. Estonia borders Russia, was part of the former Soviet Union, and has a substantial Russian-speaking minority. These factors give Russia a significant incentive to seek influence in the country. Russia has opposed Estonia's policy of Euro-Atlantic integration and increasingly has sought to undermine and delegitimize Estonia's government.

Russian activities in Estonia have been the topic of considerable study. By examining perceived Russian aggression during the Bronze Night incident of 2007, and by surveying the threat to Estonia in light of the Russian capabilities described above, this section offers a new understanding of how Russia undertakes political warfare in practice.

Historical Context

Russia has continuing influence and interest in Estonia, even as the Baltic nation has sought to integrate itself into Western institutions. Estonia became a Soviet republic following a particularly traumatic experience during World War II: Estonia was occupied by the Soviet Union first in 1939 following the signing of the Molotov-Ribbentrop pact, then captured by Nazi Germany in 1941, and then captured by the Soviet army in 1944. Many Estonians perceived the Soviet period as a foreign occupation.[50] The memory of the occupation of World War II is deeply embedded in Estonia's national history. Estonian leaders emphasize that their history guides their desire to resist Russian aggression in the future, and the country has adopted concepts of "total defence" against a potential invasion in which the entire society will be mobilized (discussions with Estonian officials, 2015). With the fall of the Soviet Union, Estonia and the other Baltic states committed themselves to Euro-Atlantic integration, and joined the EU and NATO in 2004.

During the Soviet period, many Russian speakers, including both ethnic Russians and Russian speakers from other Soviet republics, migrated to Estonia. Indeed, in 1989, the country was only 62 percent ethnic Estonians (Kasekamp, 2010). Though many Soviet citizens left after the fall of the Soviet Union, many remained. By 2011, 25 percent of Estonians were ethnic Russians, and 30 percent identified Russian as their mother tongue.[51]

From the early 1990s, in part with the goal of solidifying Estonia as a newly independent country linked with the Estonian people,

---

[50] See, for example, recent calls for reparations from Russia for its experience.

[51] For statistics, see Statistical Office of Estonia, Central Statistical Bureau of Latvia and Statistics Lithuania, 2015, p. 24.

Estonia adopted a policy of legal continuity with pre-1940 Estonia, which meant that only individuals who could trace their ancestry to before 1940 were granted Estonian citizenship. Though there has been some liberalization of Estonia's citizenship policy (Winnerstig, ed., 2014), and many Russian speakers were naturalized, today about half of the Russian speakers in Estonia are not Estonian citizens. In 2011, approximately 54 percent of ethnic Russians were Estonian citizens, 24 percent held Russian citizenship, while 21 percent had "undetermined" citizenship, which generally meant that they had been issued "Alien" papers that permit work and visa-free travel throughout the EU.[52] As shown in Figure 3.4, Russian speakers in Estonia are largely concentrated in two parts of the country: the capital, Tallinn, and Ida-Viru County, next to the Russian border.

Russian speakers in Estonia have consistently voiced frustration with their situation. Despite Estonia's advanced social welfare system, poverty and unemployment rates are higher in rural Russian majority areas (according to the 2011 census, unemployment in the country as a whole was 5.3 percent, compared with 8.3 percent in Ida-Viru County).[53] Further, Estonia's Russians cite their ethnic origin as a top basis for discrimination and as an impediment to their workplace advancement.[54] Many Russian speakers see integration and language policies as forced assimilation and feel they are often viewed as "others."

Nevertheless, for several reasons, the receptivity of Russian speakers to Russian propaganda, manipulation, and recruitment is limited. One report identifies that about 37 percent of Russian speakers are either "successfully integrated" or are "Russian speaking patriot[s]" (Kivirähk, 2014). Though lower than the rest of Estonia, average incomes in Ida-Viru are higher than in the comparable Pskov Oblast in

52 See the results of the 2011 Census in Estonia, question PC0442 (Estonia Statistics Database, 2016).

53 See 2011 Census in Estonia, question PC007 (Estonia Statistics Database, 2016).

54 One 2009 survey found that "17% of Russians in Estonia could recall an incident from the last 12 months that they considered discriminatory based on their ethnicity" (EU Fundamental Rights Agency, 2010, p. 176).

Figure 3.4
Concentrations of Russian-Speakers in Estonia

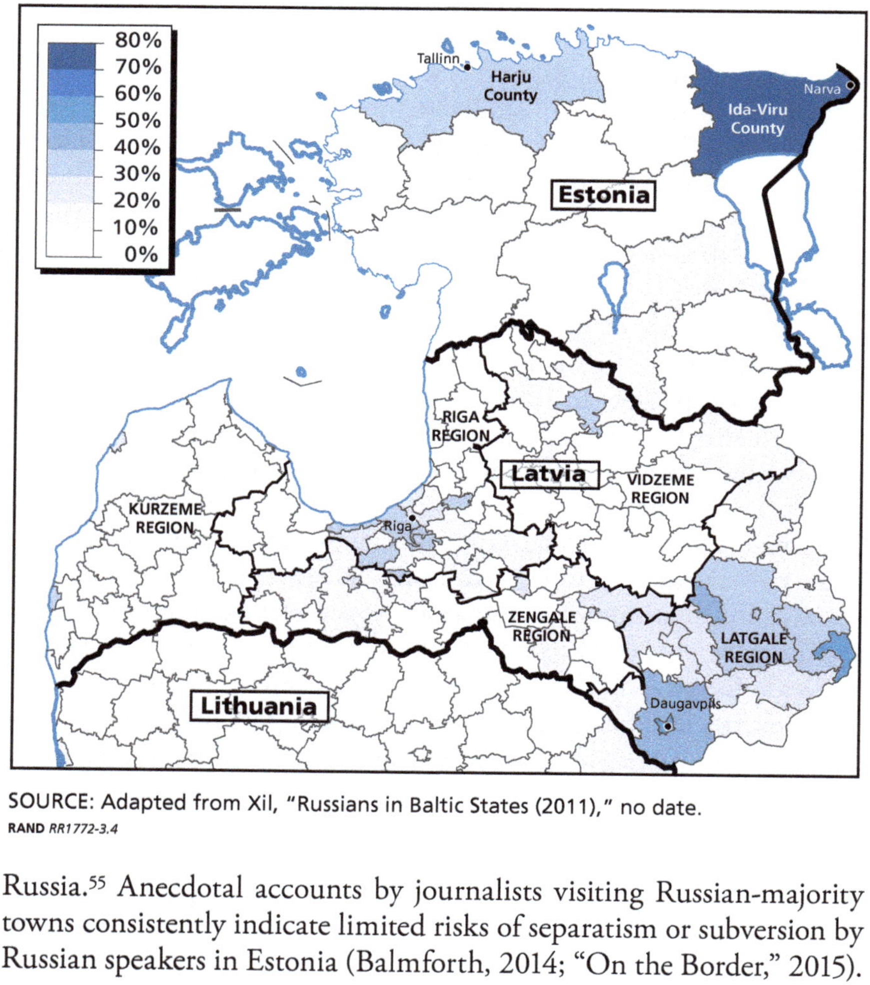

SOURCE: Adapted from Xil, "Russians in Baltic States (2011)," no date.
RAND *RR1772-3.4*

Russia.[55] Anecdotal accounts by journalists visiting Russian-majority towns consistently indicate limited risks of separatism or subversion by Russian speakers in Estonia (Balmforth, 2014; "On the Border," 2015).

---

[55] GDP per capita in Ida-Viru is $12,974 compared with $5,349 in Pskov Oblast, though Leningrad Oblast, bordering Estonia, did have high GDP per capita at $12,163. Russia in general also has higher inequality (see Estonia Statistics Database; Knoema, "Leningrad Region—Gross Regional Product Per Capita"; Knoema, "Pskov Region—Gross Regional Product Per Capita"; Internal Revenue Service (n.d.); for inequality data, see World Bank, "GINI Index [World Bank estimate]").

Estonian officials and analysts argue that Russia is seeking to exploit the Russian minority's grievances as a means of exercising influence over the country. Nevertheless, Estonian officials downplay the dissatisfaction of Russian speakers, and claim that "our Russians" are not unhappy and would not rebel against the government (interviews with Estonian officials and think tank analysts, 2015).

Accounts of Russian involvement in Estonia, especially from Estonian sources, should be seen in the context of Estonia's nation-building efforts since (re)independence. Since 1991, Estonia, as with the other Baltic states, has sought to establish an identity as an independent country reflecting the majority Estonian ethnicity. The goal of maintaining Estonian independence and identity in the country, however, has led to two contradictory trends: on one hand, Estonia recognizes a serious threat from Russia and resents Russian manipulation; while on the other hand, Estonian officials do not want to call attention to Russian grievances and thereby throw into question Estonia's success at integration.

The "Bronze Night" as an Example of Russian Political Warfare

The "Bronze Night" incidents offer a concrete example of Russian political warfare. The events began in April 2007 after Estonian authorities decided to move a statue commemorating the Soviet victory in World War II. Russian speakers and pro-Russian NGOs quickly organized to protest the statue's removal, escalating into riots and leading to one death, 150 injuries, and the arrest of more than 1,000 people (Kaiser, 2012). A series of cyber attacks followed the protests. Russia is commonly viewed at minimum as the perpetrator of the cyber attacks,[56] and in some accounts as responsible for the overall crisis ("Russia Accused of Unleashing Cyber War," 2007). Maigre describes the "Bronze Night" as "a conflict of a hybrid nature," including "riots in Tallinn, a siege of the Estonian Embassy in Moscow by pro-Kremlin Nashi youth organization demonstrators, strong economic measures imposed by Russia

---

[56] One article argues, "Russia-based botnet controllers violated Estonia's sovereignty by interfering with the nation's commerce, governance, and communication; the Russian government should have been held accountable to stop the DDOS" (Carey, 2013).

against Estonia, waves of cyber-attacks against the Estonian government and banking systems, and a fiery official Russian response" (Maigre, 2015). Similarly, the Estonian think tank ICDS has outlined purported examples of Moscow's involvement "in Tallinn's 'crystal night,'" such as the direct participation of the Russian embassy and support by Russia authorities for the protest (ICDS, 2007). The implication is that Russia has the ability and intent to use tactics across a range of domains to achieve its political objectives.

This section examines the Bronze Night incident to understand Russian political warfare in general. Contrary to the description in some Estonian sources, our analysis implies that Russia did not directly instigate the crisis. Rather, Russian political warfare activities laid the groundwork that enabled the crisis to occur, reacted in support of the protests, and likely opportunistically escalated the conflict.

What Happened?

The Bronze Soldier was a Soviet monument, erected in 1947, over the reburied remains of 12 Red Army soldiers in downtown Tallinn. World War II monuments in Estonia are highly politicized.[57] Many Estonians view Soviet monuments as a reminder of the Soviet occupation. For Russia and many Russian speakers, though, victory over the Nazis is a central element of nationalist sentiment, and many blame Estonia for collaboration with the Nazis. Russian media coverage prior to the crisis, for example, highlighted an effort by Estonian veterans groups and nationalists to erect a statue commemorating the Estonian SS Legion, though the Estonian government quickly acted to remove that statue (Brüggemann and Kasekamp, 2008, p. 448).

The Estonian government appears to have decided to move the Bronze Soldier in late 2006 to early 2007. The Estonian think tank ICDS explains the decision as the result of increasing provocation by individuals visiting the monument, including an incident on May 9,

---

[57] Karsten Brüggemann and Andres Kasekamp note that the construction of the monument was "a highly political act from the start. No fighting had taken place anywhere nearby. It was simply a prominent central location suitable for following the standard approach in other major cities of the Soviet Union of constructing a memorial with common graves that could be used for ceremonial purposes (Brüggemann and Kasekamp, 2008, p. 443).

2006, "when the police had to intervene in a conflict that arose [when] persons carrying flags of the USSR assaulted persons carrying the flag of the Republic of Estonia" (ICDS, 2007). In the run-up to elections in 2007, a number of Estonian political parties raised the issue of the presence of the Bronze Soldier in downtown Tallinn as a symbol of Soviet occupation. Robert Kaiser highlights the role of the Reform Party, led by Andrus Ansip, who "used the monument's removal as a means of mobilizing support for itself."[58] With Ansip's urging, the Estonian parliament passed two laws justifying the removal of the statue, though one law prohibiting Soviet-era symbols in public displays was eventually vetoed. Partly based on its anti-Soviet rhetoric, the Reform Party did well in the March 2007 elections, joined the new coalition, and began efforts to move the statue as promised in its campaign (see also Smith, 2008, p. 426; Brüggemann and Kasekamp, 2008, p. 435).

On April 26, 2007, the monument was fenced off and covered, so that archeologists could examine the site to determine how many bodies were present. Brüggemann and Kasekamp write, "Throughout the day a crowd of over 1,000 mainly ethnic Russians gathered. In the evening riot police managed to disperse the crowd, but the protesters now headed into central Tallinn where cars were set ablaze, windows smashed and shops looted. One young man, a Russian citizen, involved in the looting was stabbed to death" (Brüggemann and Kasekamp, 2008, p. 436). On April 27, the government decided to move the statue to a safe location, and in Moscow, the Russian nationalist youth group Nashi surrounded the Estonian embassy.

Late on April 27, a series of cyber attacks began. The first phase of cyber attacks from April 27 to 29 were relatively technologically unsophisticated, leading analysts to attribute them to Russian nationalists unconnected to the government. One report described the attacks as "emotionally motivated, as the attacks were relatively simple and any

---

58 Kaiser quotes Ansip, explaining: "I see the solution to this problem in the relocation of the monument to the cemetery. . . . It has become all the more clear that the monument cannot remain in its old place. The question rose: whose word has authority in Estonia? The word coming from the Kremlin or the word from Old Town? We cannot say to our people, that Estonia is after all only a union republic, and our word in this country is not worth a 'brass farthing'" (Kaiser, 2012, pp. 1051–52).

coordination mainly occurred on an ad hoc basis."[59] During a second phase, from April 30 to May 18, the cyber attacks dramatically increased in intensity and sophistication. A large botnet of approximately 85,000 computers was used to conduct distributed denial of service (DDoS) attacks, official websites (including that of the Reform Party) were hacked and defaced, and domain name system (DNS) servers were attacked, temporarily disrupting Internet service in certain parts of the country. E-banking services for two banks controlling approximately 75–80 percent of the Estonian market were temporarily shut down for a period of hours on two occasions (Tikk, Kaska, and Vihul, 2010, pp. 20–22). Though the attack was clearly significant to Estonia, one U.S. analyst observed that the attack "from a technical standpoint is not something we would consider significant in scale" (quoted in Heickerö, 2010, p. 42).

Russian politicians, groups, and the government responded to the crisis with strong opposition to Estonia's decision to move the statue (Myers, 2007). The lower house of the State Duma unanimously passed a resolution accusing the Estonian parliament of "glorifying fascism" and urging President Putin to consider imposing economic sanctions. The Duma's Committee on Foreign Relations proposed that Russia introduce economic sanctions and highlighted Estonia's energy dependence on Russia. Russia's Federal Council called to sever diplomatic ties with Estonia (Rozhkov ☒☒☒☒☒☒, 2007). Russia's Minister of Foreign Affairs, Sergei Lavrov, called the events in Estonia "disgusting" and noted that they would have "serious consequences for our relationships with Estonia" ("Tallin," 2007"). In retrospect, two years later, Lavrov would justify Russia's economic response by observing that the Russian public reacted with outrage to the plan to move the memorial. "Most of our compatriots understood and supported these actions," he said ("Lavrov," 2009).

Russian companies followed government rhetoric by effectively sanctioning Estonia, though they generally justified their actions in

---

[59] The initial attack was apparently begun when instructions for executing ping commands was posted on various Russian-language Internet sites. One paper explains, "As a generalisation, though, the initial attacks on April 27 and 28 were simple, ineptly coordinated and easily mitigated" (Tikk, Kaska, and Vihul, 2010, p. 18).

other ways. The Russian Rail Company, RZhD—which enjoys a monopoly on all rail trade with Russia—stopped oil deliveries to Estonia for several weeks following the incident, claiming the need for urgent repairs. According to Estonia's Rail Company, freight volume decreased by 35 percent in 2007 compared with 2006, raw oil by 34 percent, petroleum by 50 percent, and coal by more than 60 percent ("Ssora s Rossiej obernulas' dlja Estonii ogromnymi poterjami" [Fight with Russia Brought Estonia Great Losses], 2007). Further, many Russian companies refused to buy Estonian products.[60] The North Lead Alliance delayed construction of a factory in Estonia, the Akron investment company refused to finance projects in Estonia, and a chain of stores in Bashkortostan stopped selling Estonian goods (Starikov, n.d.). Authorities in the St. Petersburg region severely limited transportation through the Narva River bridge that connects Russia and Estonia.[61] The Russian newspaper *Kommersant* quoted an Estonian newspaper reflecting on the crisis: "A year after the scandalous incident with the moving of bronze Alyosha' the government counted its losses: €450 million, or 3 percent of the GDP. The Port of Tallinn especially feels its losses. It lost 13 percent of its transit turnover" ("Tallinn," 2008).

Analysis

Estonian accounts of the crisis highlight four reasons to attribute the crisis to Russia. First, they note Russia's influence in sustaining a narrative opposed to Estonia's view of World War II. Second, they claim there was direct support and communication between the Russian embassy and the "Night Watch," a youth group (allegedly supported by Russia) that appears to have played a significant role in rallying the rioters. The Night Watch was apparently established "to protect the monument from being profaned by provocateurs and vandals," and held a number of protests against the removal of the statue prior to the

60 For example, Estonia's Kalev chocolate factory was subject to a boycott (see http://psy.rin.ru/cgi-bin/news.pl?id=124331).

61 http://www.vesti.ru/doc.html?id=114941&cid=5; Similarly, a member of the United Russia's committee responsible for work with the displaced persons and compatriots abroad proclaimed that "boycotting Estonian products—is the least of measures that Russia should take in relation to Estonia" (http://www.niann.ru/?id=315506).

Bronze Night ("A Brief History," 2007).[62] Third, the protests of the Estonian embassy in Moscow, by Nashi, were said to be organized by the regime. Fourth, Russian "cyber militia" or other state-controlled actors were believed to be responsible for the cyber attacks (Carey, "Cyber Militias," 2013).

Russia did take a number of actions to expand its influence in Estonia, which likely did contribute to the protests. Furthermore, its own narrative of World War II may have set the stage for a social clash. However, scarce evidence exists that Russia initiated and executed the entire crisis. Post-hoc Russian support for the protests is not evidence of Russia's intention to start a crisis.

Claims of Russian responsibility appear to be based primarily on the role of the Night Watch and alleged Russian government responsibility for the cyber attacks. The riots rapidly escalated on April 26, and even if Russian officials did consult with some protestors, it is difficult to believe, given the rapid growth and large scale of the protests, that they were single-handedly coordinated by Russia. Reports at the time from an Estonian official reinforce the idea that the riots were spontaneous rather than planned or carefully organized.[63] Similarly, it appears that the initial cyber attacks were executed by nongovernmental actors. It is theoretically possible that the Russian government decided to begin its attack by appearing unsophisticated, though this seems an overly complex explanation. Though there is extensive discussion of other elements of the Bronze Night incident in Russian media, there is virtually no talk or analysis of Russia's role in the cyber attacks in Estonia or in mobilizing protest movements there, except for Putin's press secretary's denial that Russia had anything to do with the attacks.[64]

---

[62] Nevertheless, an Estonian court acquitted the Night Watch of charges of riot instigation (see Sputnik, 2009).

[63] One report notes, "According to Interior Minister Juri Pihl, speaking after the second night, the rioting did not seem to reflect organized preparations, planning, or clear leadership, but for the most part a mob spirit and spontaneous dynamic" (Socor, 2007).

[64] The most concrete discussion was from Sergei Markov, a State Duma deputy from United Russia, who likely speaking in jest identified his assistant as the perpetrator, "About the cyberattack on Estonia . . . don't worry, that attack was carried out by my assistant. I won't tell you his name, because then he might not be able to get visas" (Coalson, 2009).

Estonia attempted to investigate and prosecute those responsible, with little success.[65] Analysis of the events tends to highlight the absence of concrete evidence of Russian instigation,[66] though many U.S. and Estonian sources tend nevertheless to hold Russia responsible.[67] It is more likely that the Russian state played a role in supporting the cyber attacks during the second wave.

This review of events suggests a more interactive and opportunistic model for Russian political warfare, with Russia sowing seeds for conflict in fertile conditions and escalating the resulting unrest. Significant opposition to Estonian activities by local Russian speakers led to protests and cyber attacks by relatively untrained individuals. The Russian authorities then capitalized on the opportunity to oppose, undermine, and attack the Estonian government by imposing sanctions

---

65 Cyrus Farviar writes, "Partly because of the evidence that he's seen, and Moscow's reluctance to be cooperative leads made Estonian Chief Prosecutor Margus Kurm say that he is confident that the leaders of the attacks are in Russia, despite saying: "We have no evidence and no information that this was the Russian government." Still, Kurm is pretty hopeless of ever gaining any further information that could be legally useful for prosecuting anyone for cybercrimes against the Republic of Estonia. In an interview in July 2007, he admitted to me: "The status is that we haven't got any information from Russia and I'm quite sure that we will not get any information" (Farivar, 2007).

66 One article observed, "The Russian government denied responsibility, and outside experts found the evidence of official government involvement weak. Mikko Hypponen, chief research officer at security company F-Secure, feels that the attacks would've been more effective if the Russian government had been involved. The maximum bandwidth used was roughly 90 Mbps, with 10 attacks lasting 10 hours or more. . . . By comparison, in May 2006, the antispam company Blue Security was forced to shut down by a retaliating spammer who used tens of thousands of machines in a DDoS attack amounting to 2 to 10 Gbps of fake traffic" (Lesk, 2007).

67 A CNAS report explained: "In 2007, Estonia was deluged by cyber attacks. All signs pointed to Russian involvement: Many of the cyber attacks themselves were traced to Russia; many of the attack tools were written in Russian; many of the corrupted Estonian websites were polluted with strong nationalist Russian reactions; numerous Russian politicians openly supported the attacks; and the Russian government refused to stop or even investigate the attacks. Russian police simply remained on the sidelines." Nevertheless, these activities could be expected if Russia supported the attacks but did not initiate them. Similarly, Stephen Herzog observed, "While we may never know the true extent of Kremlin involvement in the cyber attacks on Estonia, it is clear that Russian officials encouraged the hackers by accusing Tallinn of altering history, perpetrating human rights violations, and encouraging fascism" (Denmark and Mulvenon, 2010, p. 170; Herzog, 2011, p. 53).

against Estonia; encouraging the activities of its proxies, including the directly controlled Nashi; and escalating the cyber attacks. Estonian officials emphasize that Russia retains the ability to begin a protest through its networks in Estonian society (discussions with Estonian officials and think tank analysts, 2015). While this may be true, the size and legitimacy of protests and the severity of the unrest depends on the overall environment as much as on Russian actions.

Current Russian Means of Influence in Estonia

Estonian government agencies are deeply concerned with the ongoing and cumulative effects of Russia's activities to destabilize the country, Merle Maigre, an advisor to the Estonian president, notes that there is "nothing new" in Russia's tactics in Estonia (Maigre, 2015). KAPO compiles an extensive dossier of Russia's subversion efforts in its annual reports. The 2013 report explained that "Russia wants to expand the Russian empire using Russian-speaking residents as one of its tools. The need to protect Russian people is used as an excuse for aggression, although it clearly ignores facts that the entire world can see and hear through the free media" (Estonian Internal Security Service [KAPO], 2013, p. 2). Perceived Russian activities in Estonia can be roughly divided into the same categories as in the previous section.

Diplomatic and Proxy

A wide range of pro-Russian organizations in Estonia have varied links to the Russian government. These organizations may simultaneously undertake actions with no apparent coordination.

Some category 2 organizations are formally independent but transparently funded by the Russian government, generally through Rossotrudnichestvo and the Compatriot Policy.[68] The Russian embassy in Tallinn reportedly has led a "Coordination Council of Russian Compatriots," which criticized Estonia during the Bronze Night incident

[68] Specifically, the "The Fund for the Legal Protection and Support of Russian Federation Compatriots Living Abroad" (see Winnerstig, 2014, p. 40).

for breaking up "antifascist clashes" (Winnerstig, 2014, p. 43). The Compatriot Policy also funds a "Russian Ombudsman," who claims to represent the interests of ethnic Russians (Winnerstig, 2014, p. 41). Russian backed groups include the "MTÜ Impressum," a "media club" hosting Russian media figures in Tallinn (Estonian Internal Security Service [Kaitsepolitsei], 2013, pp. 8–9). Russia also sponsors youth camps through Rossotrudnichestvo, which KAPO claims are closely connected to the GRU (Estonian Internal Security Service [Kaitsepolitsei], 2013, pp. 8–11).

Some category 3 organizations, while they claim to be independent, are likely proxies of the Russian regime. These include "anti-fascist" organizations in Estonia, such as Mir Bez Natsizma (World without Nazism [MBN]). An umbrella organization present throughout the post-Soviet world, MBN claims to have 14 member organizations from Estonia, including the Night Watch. These 14 organizations are likely small, with overlapping membership, perhaps indicative more of Russia's tendency to create fake or virtual organizations than a real differentiation of opinion among Russian speakers in Estonia. The Night Wolves, the biker gang linked to Putin, also appear to be active in Estonia and seeking to expand.[69] Estonian news sources noted concern about the Night Wolves' plans and suspicion at efforts by the Night Wolves to transport embargoed supplies from Estonia to the separatists in the Donbas ("Russian Bikers Help Estonia to Send Humanitarian Aid to Donbass," n.d.).[70] Some representatives from Russian-speaking Estonian organizations welcomed the Night Wolves, while others expressed concern about the Night Wolves "con[s]orting with authorities and scaring young people" and bringing unwanted hostility

69 Galleotti describes a potential scenario in which "the Kremlin-backed Night Wolves motorcycle gang tries to force the border into Narva, unarmed but in numbers," complementing events such as protests by Russian speakers in Narva or bombs in Tallinn (Galeotti, 2015).

70 The leader of the volunteer group Dobrosvet near Tallinn offers a good example of the loyalty of many Russian speakers to Estonia—though her group collects humanitarian aid for Night Wolves to ship to the Donbas through Russia, her son will serve in the Estonian army—she notes, "Russians in Estonia 'already have a different mentality.'" (See "On the Border," 2015).

in Estonia ("Moscow Bikers' Club with Links to Putin May Expand to Estonia," 2013).

Some category 4 organizations have independent decisionmaking authority but may share an agenda with the Russian regime. These organizations tend to have constituencies and legitimate concerns for the Russian-speaking population. The Legal Information Centre for Human Rights (LICHR), for example, cooperates with Amnesty International and has criticized Estonia's policy toward the Russian population. The KAPO has noted that funding for LICHR comes from Russia (Winnerstig, 2014, p. 47; Braw, 2015; Poleshchuk, 2014). Other groups with a similarly declared agenda include the "Russkaja Shkola Estonii" (Russian School of Estonia), which backs Russian language schools, and the Estonian Aliens Union, which seeks support from Western leaders on citizenship issues ("Union of Non-Citizens of Estonia").

The Centre Party in Estonia, which receives most of its support from Russian speakers and was the second largest party in the last elections, also falls in category 4.[71] Despite its size, the Centre Party has not been included in political coalitions, in part because of concern that it might be influenced by Russia ("On the Border," 2015; discussions with Estonian officials and think tank analysts, 2015). Furthermore, the KAPO accused Edgar Savisaar, the leader of the Centre Party, of seeking money from Vladimir Yakunin, the head of the Russian Railway Company, to build an Orthodox cathedral, precipitating the "Eastern Money Scandal" of 2010. Vadim Polescchuk, of the pro-Russian LICHR, countered that the KAPO appeared to have inappropriately listened in on the discussions of the leader of an opposition political party (Savisaar), the Estonian media had unfairly accused Yakunin of being a member of the KGB, and the Orthodox Church was not controlled by the regime (Poleshchuk, 2014, p. 50). Nevertheless, the Centre Party is believed to have a formal agreement with United Russia, presumably in exchange for funding. Russian media

[71] According to the LICHR, in 2011 the Centre Party was "supported by 81 percent of Russian-speaking citizens and 11 percent of enfranchised Estonians," who are generally lower-income and older Estonians (Poleshchuk, 2014, p. 14, 17; discussions with Estonian official, 2015).

also shows positive programming toward Centre Party figures (FOI, 2010, pp. 50–51; Grigas, 2012; Tagel, 2015).

Informational

Russian speakers are often described as living in a "different information space" from Estonians. Russian speakers watch TV stations broadcast primarily from Russia, including PBK, a Russian-language station rebroadcasting Russian media;[72] Rossia/RTR, a widely watched Russian station; NTV, which is owned by Gazprom; and RTV, broadcast by RT (Kaljurand 2015; Loit, n.d). Analysts and observers explain the preferences of Russian speakers for Russian language media as being due to better content. Many of the stations have high-quality Western programs (e.g., *Game of Thrones*), occasionally interspersed with news. The news broadcasts, following the models of broadcasts within Russia, are generally subtle in promoting the regime's messages (Kaljurand 2015; interview with think tank analyst, 2015).

Cyber

A major vulnerability in Estonia is cyber attacks. Estonia is one of the most networked societies in Europe—the entire government operates online, partly as preparation to move the government infrastructure abroad in case of a Russian attack. In 2014, the KAPO reported an "advanced persistent threat" that repeatedly targeted Estonian government networks (Estonian Internal Security Service [Kaitsepolitsei], 2014, p. 19). While Estonian sources highlight the role of the FSB in cyber warfare, there is little information to conclusively link ongoing cyber attacks with the Russian government (Estonian Internal Security Service [Kaitsepolitsei], 2013, p. 13). Indeed, as one cyber security official in Estonia explained,

> [Russia] does not correspond to our principle of a state with separation of powers, with a functioning civil society, where organised crime also stands separate from the state. Over there, it is all

---

72 BMA, which controls a variety of stations, is co-owned by "Russian-born Latvian businessman Oleg Solodov and Russian citizen Alexei Plyasunov" (Krutaine and Sytas, 2015; Spriņģe, Motuzaite, and Gailāne, 2012).

> intertwined. Therefore, a cyber attack issued from Russia may come from a criminal organisation or a youth organisation or wherever. But eve[n] while it is often technically complicated to ascertain the source of an attack, we can always use our common sense and analytical mind: if the attack was politically motivated, there might be someone behind it with political motives (Pau, 2015).

Hence, while Estonia highlights the threat from Russia, it recognizes that it faces a wide range of potential cyber attackers with parallel agendas, even if they are not directly controlled by the Russian state.

Intelligence

Russian intelligence is believed to be active in Estonia, as shown by several cases of apparent Russian espionage. In December 2014, the FSB claimed it had an agent, Uno Puusepp, who had operated for 20 years in the KAPO ("FSB Reveals It Had Agent in Estonia Intel for 20 Years" 2014). As discussed above, according to Estonian authorities, the FSB also operated in Estonian territory when they seized KAPO agent Eston Kohver in September 2014 (interviews with Estonian officials, 2014). After being sentenced to 15 years' imprisonment by a Russian court, Kohver was exchanged in September 2015 for Aleksei Dressen, a KAPO officer convicted of working with the FSB. In 2013, the KAPO announced that Vladimir Veitman, who had been working for the KAPO since 1991, in fact had been recruited by the SVR. In 2009, Estonia caught Herman Simm, an official in the Ministry of Defence, who had shared NATO classified information with Russia (Schmid and Ulrich, 2010). KAPO reports claim the SVR, FSB, and GRU are active in Estonia (Estonian Internal Security Service [Kaitsepolitsei], 2013, pp. 13–17).

Economic

Estonia does have significant economic relationships with Russia, and Russian actions taken after the Bronze Soldier incident cost Estonia 1.85 percent of its GDP.

A significant volume of Russian goods is transported through Estonia, especially Estonian railways. An estimated 80 percent of rail traffic in Estonia is transshipment, as Estonia shares the same rail gauge as Russia. Rail traffic in Estonia dropped significantly in 2007 following the Bronze Night incident, potentially pointing to Estonia's vulnerability to Russian leverage. However, Estonian officials are actively considering changing the rail gauge to an EU standard, which would diminish Estonia's attractiveness as a transit country. This follows a continuing trend of Estonian officials insisting that they intend to deepen ties with the West over those with Russia (Schaefer, 2011; discussions with Estonian officials, 2015).

Estonia depends on Russia for some of its energy supplies, especially natural gas, but its dependence affects a relatively small portion of energy supplies, and that portion is declining. Estonia currently receives all of its natural gas supplies from Russia, and is hence vulnerable to a cutoff in supply. However, EU estimates indicate that by sharing gas with its neighbors, Estonia could replace 40 percent of its supplies (European Commission, 2014). Furthermore, natural gas represents a relatively small portion of the Estonian energy supply—only 6.4 percent of total energy consumption in 2014—and the opening of the Klaipeda LNG port in Lithuania in 2015 enabled Estonia to further diversify its supply (Eurostat, 2016; Rapoza, 2015). Estonia does not import electricity from Russia, but Latvia and Lithuania do, and they re-export some of this electricity to Estonia. Russia could theoretically cut off electricity supply to the Baltic countries, though not without also cutting off Kaliningrad. Overall dependence on Russia for electricity in the region is declining with new interconnections with the European electricity supply (Larrabee et al., 2017, pp. 31–33; Kropaite, 2015).

Finally, Russia has significant trade and investment with Estonia, though not to the point that a trade embargo alone would deeply undermine the Estonian economy. Exports to Russia constituted more than 10 percent of the export market share for Estonian agricultural, livestock, and fishery goods, hence pointing to Estonia's vulnerability to ongoing Russian countersanctions against European agricultural goods. However, these goods constitute a very small portion of Estonia's overall exports to Russia, and Estonia has been consistent in its support for sanctions against Russia (discussions with Estonian officials, 2015; Larrabee et al., 2017, pp. 31–33).

Conclusion

This chapter has examined Russian thinking and practices regarding political warfare. Russia's political warfare activities are informed by Russian understanding of new generation warfare, by Russian and Soviet historical uses of political warfare, and by recent Russian domestic politics. While Russia has wide-ranging capabilities, including proxies, nonkinetic use of military forces, information activities, cyber warfare, and economic influence, these tools are not unlimited and are generally used in specific contexts in limited and opportunistic ways. Based on the observations above, we highlight three central observations about Russian political warfare that have important implications for defense and deterrence against Russia and for adapting U.S. and allied political warfare to be more effective.

Russian Political Warfare Is a Response to Western Political Warfare

Analysis of Russia's understanding of contemporary political warfare indicates that Russia sees itself largely on the defensive. Many analysts have sought to demonstrate an alignment between political warfare activities in Ukraine and writings on new generation warfare, pointing to the adoption of political warfare as an offensive strategy.

However, our own analysis of primary sources, as well as observations from other close examinations of the Russian military discourse, emphasize that the Russian leadership sees a major deficit in its own ability to use and coordinate soft power and political warfare tools. Russian intelligence services and other actors compete and often collide in conducting their missions and are at times pursuing divergent agendas. Russia tends to overestimate the role of the United States in spurring democratic and liberal revolutions through the Maidan protests, color revolutions, Arab Spring, and so forth. The tendency of the Putin regime to see these events as artificial developments of aggressive Western policy is likely reinforced by Putin's own history in Soviet intelligence and the semi-democratic character of Russia's regime. Understanding Russia's true motivations and mind-set is critical to developing approaches that will deter rather than galvanize aggressive behavior.

Russian Political Warfare Stokes Conflict and Capitalizes on Crises

Rather than manufacturing political crises from start to finish, Russia appears to operate by creating pressure and intensifying social divides, and then taking advantage of crises once they emerge. This tendency is clearly visible in the case of the Bronze Night. Russian propaganda and narratives directed at Russian speakers set the stage for the denouement of the crisis, and Russia provided financial and other support to groups that played important roles, the Night Watch being the best example. But Russia does not appear to have initiated the crisis, and its activity in the crisis appears to be reactive. Similarly, in eastern Ukraine, though the entire story is not yet fully clear, it seems that Russia supported counter-Maidan protests by Yanukovich's Party of Regions and backed propaganda claiming that the Maidan protesters were fascists (Meek, 2014). Once the protest movement grew with limited Russian support, Strelkov, Borodai, and others, possibly at the urging of the Russian leadership, encouraged the shift into a separatist movement, taking advantage of economic grievances and the close links between Russia and the Donbas (Zhukov, 2014). Again, Russia reacted to a crisis that emerged out of tension that it had helped create.

This opportunistic mode of operation is consistent with Russian descriptions of Putin's strategic thinking. In a public lecture, Lukyanov described Putin as "not a strategic thinker"—rather, Putin is "prepared for any development" and ready for "immediate reaction" (Lukyanov, 2016). Similarly, Sergei Pugachev, a former aide, noted, "Putin is not someone who sets strategic plans; he lives today" (Bullough, 2014). Monitoring and tracing Russia's effort to feed crises may therefore be critical for U.S. and NATO efforts to counter Russian political warfare. Understanding financial flows and the ways in which Russia's network of loosely affiliated proxies and organizations operate is also essential.

Russian Political Warfare Depends on the Geographic and Political Context

Russia's use of political warfare is constrained and determined in part by the geographic and political context. Its capabilities and influence appear noticeably greater within the post-Soviet world and its near abroad. For example, the "Regional Priorities" section of Russia's Foreign Policy Concept identifies, in decreasing specificity, goals for the CIS Member States, neighboring countries such as Georgia, the EU and NATO, the United States, the Far East, and the rest of the world. The FSB, with primary responsibility for Russia and the near abroad, has significantly greater resources than the SVR, which has greater responsibility for Western Europe and the United States. The FSB also relies on preexisting networks from the Soviet period, implying that it is more difficult, though certainly not impossible, for the FSB to operate effectively beyond the post-Soviet world.[73] Russia's media and information activities also seem significantly more effective in Russian-speaking areas than in Western Europe. Finally, surveys and research on Estonia, eastern Ukraine, and other parts of Russia's near abroad emphasize that while some individuals remain close to Russia and are quite responsive to Russian propaganda, others are increasingly integrated into the West and correspondingly less responsive (Kivirähk, 2014; Pifer and Thoburn, 2014). Russia certainly can exert influence outside of its near abroad, as it has in Syria. Nevertheless, Russia's influence and leverage are likely to be proportionately greater in its region, given factors including the presence of Russian speakers, preexisting linkages with the target society, low economic development, and a shared Soviet narrative.

References—Chapter 3

Abdelal, Rawi, "The Profits of Power: Commerce and Realpolitik in Eurasia," *Review of International Political Economy*, Vol. 20, No. 3, 2013, pp. 421–56.

"Actions to Promote Discord. Folder 90. The Chekist Anthology," June 2007, History and Public Policy Program Digital Archive, Contributed to CWIHP by Vasili Mitrokhin. As of April 24, 2016:
http://digitalarchive.wilsoncenter.org/document/113684

[73] In the case of the poisoning of Litvinenko, for example, the FSB's efforts have been largely revealed and demonstrate for example clumsy handling of the radioactive material (*BBC News*, 2016; discussions with think tank analysts, 2015–February 2016).

Adamsky, Dima, "Cross-Domain Coercion," Institut Français des Relations Internationales, November 2015. As of April 25, 2016:
http://www.ifri.org/sites/default/files/atoms/files/pp54adamsky.pdf

"Alexander Litvinenko: Russian Spy," BBC News, January 21, 2016. As of April 24, 2016:
http://www.bbc.com/news/uk-19647226

Alpervotich, Dmitri, "Bears in the Midst: Intrusion into the Democratic National Committee," June 15, 2016. As of August 2, 2016:
https://www.crowdstrike.com/blog/bears-midst-intrusion-democratic-national-committee/

"Analysis of Russia's Information Campaign against Ukraine," NATO Strategic Communication Center of Excellence, 2014.

Armstrong, Matt, "False Rivals: How RT Is Larger & Targets Different Audiences Than the BBG," May 19, 2015. As of April 24, 2016:
http://www.mountainrunner.us/blog/2015/5/false-rivals-how-rt-competes-with-commercial-journalism-not-the-bbg

"Association of the United Postwar Immigrants. Folder 52. The Chekist Anthology," June 2007, History and Public Policy Program Digital Archive, Contributed to CWIHP by Vasili Mitrokhin. As of April 24, 2016:
http://digitalarchive.wilsoncenter.org/document/113609

Balmforth, Tom, "Russians of Narva Not Seeking 'Liberation' by Moscow," RFE/RL, April 4, 2014. As of April 24, 2016:
http://www.rferl.org/content/russia-estonia-notcrimea/25321328.html

Baluyevskiy, Yuri, and Musa Khamzatov, "Globalizacija i voennoe delo" [Globalization and Military Affairs], 2014. As of April 24, 2016:
http://nvo.ng.ru/concepts/2014-08-08/1_globalisation.html

Bartles, Charles, and Roger McDermott, "Russia's Military Operation in Crimea," *Problems of Post-Communism*, Vol. 61, No. 6, 2014, pp. 51–52.

BBC News, "Russian-language Media in the Baltic States," August 24, 2015. As of April 25, 2016:
http://www.bbc.co.uk/monitoring/russianlanguage-media-in-the-baltics; See also discussions at http://www.diplomaatia.ee/en/article/estonias-virtual-russian-world-the-influence-of-russian-media-on-estonias-russian-speakers-1/

———, "Ukraine Crisis: Key Players in Eastern Unrest," August 28, 2014. As of April 24, 2016:
http://www.bbc.com/news/world-latin-america-27211501

Bechev, Dimitar, *Russia's Influence in Bulgaria, New Direction: The Foundation for European Reform*, n.d. As of July 14, 2016:
http://europeanreform.org/index.php/site/publications-article/russias-influence-in-bulgaria

Berzina, Ieva, "Zapad 2013 as a Form of Strategic Communication," in Zdanavičius and Czekaj, eds., *Russia's Zapad 2013 Military Exercise: Lessons for Baltic Regional Security*, Washington, D.C.: Jamestown Foundation, 2015, pp. 65–69.

Berzins, Janis, "Russia's New Generation Warfare in Ukraine: Implications for Latvian Defense Policy," National Defence Academy of Latvia Center for Security and Strategic Research, Policy Paper No. 2, April 2014.

Bidder, Benjamin, "Russia Today: Putin's Weapon in the War of Images," *Spiegel Online International*, August 13, 2015. As of April 24, 2016:
http://www.spiegel.de/international/business/putin–fights-war-of-images-and-propaganda-with-russia-today-channel-a-916162.html

Blackwill, Robert, and Jennifer Harris. *War by Other Means: Geoeconomics and Statecraft.* Cambridge, Mass.: Belknap Press of Harvard University Press, 2016.

Blank, Stephen, "Russian Information Warfare as Domestic Counterinsurgency," *American Foreign Policy Interests*, Vol. 35, No. 1, January–February 2013, pp. 31–44.

"Bojkotirovanie jestonskih tovarov—malaja chast' togo, chto rossijskoj storone neobhodimo predprinjat' v otnoshenii rukovoditelej jetoj strany," Ljubomir Tjan [Boycotting Estonian products—Is a Small Part of What Russia Should Do in Relation to the Leaders of this Country, Lyubomir Tyan], NIA Nizhniy Novgodor, May 3, 2007. Available at
http://www.niann.ru/?id=315506

Bordachev, Timofei, "Russia and the Eurasian Economic Union: The View from Moscow," European Council on Foreign Relations, January 21, 2015. As of October 4, 2016:
http://www.ecfr.eu/article/commentary_russia_and_the_eurasian_economic_union_the_view_from_moscow403

Braw, Elizabeth, "The Kremlin's Influence Game," *World Affairs Journal*, March 10, 2015. As of April 25, 2016:
http://www.worldaffairsjournal.org/blog/elisabeth-braw/kremlin's-influence-game

"A Brief History of the Night Watch in Estonia," May 7, 2007. As of April 24, 2016:
http://www.cafebabel.co.uk/tallinn/article/brief-history-of-nightwatch-in-estonia.html

Broadcasters' Audience Research Board (BARB), *BARB Weekly Viewing Summary.* As of April 24, 2016:
http://www.barb.co.uk/whats-new/weekly-viewing-summary?period_year%5B%5D=2015&period_month%5B%5D=9&period_week%5B%5D=6&button_submit=View+figures&period%5B%5D=201509060106

Brüggemann, Karsten, and Andres Kasekamp, "The Politics of History and the 'War of Monuments' in Estonia," *Nationalities Papers*, Vol. 36, No. 3, 2008, p. 443.

Geiger, Friedrich, "German Exports to Russia Fall off Further as Sanctions Take Hold: New Data Shows Sharp Decline," *Wall Street Journal*, October 29, 2014.

Gerasimov, Valery, "The Value of Science in Prediction," *Military-Industrial Kurier,* February 27, 2013, trans. Robert Coalson. As of April 25, 2016: http://www.huffingtonpost.com/robert-coalson/valery-gerasimov-putin-ukraine_b_5748480.html

Glaser, April, "Here's What We Know about Russia and the DNC Hack," *Wired,* July 27, 2016. As of August 2, 2016: https://www.wired.com/2016/07/heres-know-russia-dnc-hack/

Goodin, Dan, "First Known Hacker-Caused Power Outage Signals Troubling Escalation," *ars technica*, January 4, 2016. As of April 24, 2016: http://arstechnica.com/security/2016/01/first-known-hacker-caused-power-outage-signals-troubling-escalation/

Gorenburg, Dmitry, "Russian Military Reform: Tracking Developments in the Russian Military," Moscow Conference on International Security 2014, part 1: The Plenary Speeches, May 29, 2014. As of April 24, 2016: https://russiamil.wordpress.com/2014/05/29/moscow-conference-on-international-security-2014-part-1-the-plenary-speeches/

Grigas, Agnia, "Legacies, Coercion, and Soft Power," Chatham House, August 2012, As of January 29, 2017, https://www.chathamhouse.org/publications/papers/view/185321

Heickerö, Roland, "Emerging Cyber Threats and Russian Views on Information Warfare and Information Operations," *FOI*, March 2010.

Herszenhorn, David, "Facing Russian Threat, Ukraine Halts Plans for Deals with E.U.," *New York Times,* November 21, 2013. As of April 24, 2016: http://www.nytimes.com/2013/11/22/world/europe/ukraine-refuses-to-free-ex-leader-raising-concerns-over-eu-talks.html?_r=0

———, "Russia Putting a Strong Arm on Neighbors," *The New York Times,* October 23, 2013. As of April 24, 2016: http://www.nytimes.com/2013/10/23/world/europe/russia-putting-a-strong-arm-on-neighbors.html

Herzog, Stephen, "Revisiting the Estonian Cyber Attacks: Digital Threats and Multinational Responses" *Journal of Strategic Studies*, Vol. 4, No. 2, Summer 2011, p. 53.

Hill, Fiona and Clifford Gaddy, *Mr. Putin: Operative in the Kremlin,* Washington, D.C.: Brookings Institution Press, 2013.

"How Americans Access Their News," American Press Institute, March 17, 2014. As of April 24, 2016: https://www.americanpressinstitute.org/publications/reports/survey-research/how-americans-get-news/

"How to Deal with Harmony," *The Economist*, October 5, 2014. As of October 24, 2016:
http://www.economist.com/blogs/easternapproaches/2014/10/latviaselection?zid=307&ah=5e80419d1bc9821ebe173f4f0f060a07

Hurt, Martin, "Lessons Identified in Crimea: Does Estonia's National Defence Model Meet Our Needs?" April 2014. As of April 25, 2016:
http://www.icds.ee/fileadmin/media/icds.ee/failid/Martin%20Hurt%20-%20Lessons%20Identified%20in%20Crimea.pdf

ICDS, "Russia's Involvement in the Tallinn Disturbances," May 11, 2007. As of April 24, 2016:
http://www.icds.ee/publications/article/russias-involvement-in-the-tallinn-disturbances/

Ideon, Argo, "Expert: FSB Likely Fearing Capture of Vital Agent in Estonia," *Postimees*, December 30, 2014. As of April 24, 2016:
http://news.postimees.ee/3039997/expert-fsb-likely-fearing-capture-of-vital-agent-in-estonia

IISS, *The Military Balance 2014*, pp. 161–162.

Information Security Doctrine of the Russian Federation, 2008. As of April 24, 2016:
http://archive.mid.ru//bdomp/ns-osndoc.nsf/1e5f0de28fe77fdcc32575d900298676/2deaa9ee15ddd24bc32575d9002c442b!OpenDocument

"IntelTrak," RWR Advisory Group. As of April 24, 2016:
http://www.rwradvisory.com/about-inteltrak/

Internal Revenue Service, "Yearly Average Currency Exchange Rates." As of April 24, 2016:
https://www.irs.gov/Individuals/International-Taxpayers/Yearly-Average-Currency-Exchange-Rates

Jackson, Jasper, "RT Sanctioned by Ofcom over Series of Misleading and Biased Articles," *The Guardian*, September 21, 2015. As of April 24, 2016:
http://www.theguardian.com/media/2015/sep/21/rt-sanctioned-over-series-of-misleading-articles-by-media-watchdog

Jane's Sentinel Security Assessment, "Russia and the CIS: Russian Federation, Security and Foreign Forces," August 19, 2015. As of April 25, 2016:
https://janes.ihs.com/Janes/Display/1303056

Jones, Sam, "Photos and Roses for GRU 'Spetsnaz' Casualties," *Financial Times*, August 8, 2014. As of April 24, 2016:
http://www.ft.com/intl/cms/s/0/f28913f6-1d8f-11e4-8f0c-00144feabdc0.html#axzz42XZK1ycu

Jones, Sam, and Roman Olearchyk, "Russia Troops Build Up on Ukraine Border," *Financial Times,* August 5, 2014. As of April 24, 2016:
http://www.ft.com/intl/cms/s/0/fae9bde0-1ca2-11e4-88c3-00144feabdc0.html#axzz3xzJFbS9v

Jonsson, Oscar, and Robert Seeley, "Russian Full-Spectrum Conflict: An Appraisal after Ukraine," *Journal of Slavic Military Studies*, Vol. 28, No. 1, 2015, pp. 8, 20–21.

Kaiser, Robert, "Reassembling the Event: Estonia's 'Bronze Night,'" *Environment and Planning D: Society and Space,* 2012, Vol. 30, 2012, pp. 1051–52.

Kaljurand, Riina, "Estonia's 'Virtual Russian World': The Influence of Russian Media on Estonia's Russian Speakers," *Diplomaatia*, No. 147, November 2015.

Kasekamp, Andres, *A History of the Baltic States,* New York: Palgrave Macmillan, 2010.

Keck, Zachary, "Russia Threatens to Deploy Nuclear Weapons in Crimea," *National Interest*, June 1, 2015. As of April 24, 2016:
http://nationalinterest.org/blog/the-buzz/russia-threatens-deploy-nuclear-weapons-crimea-13013

Kennedy, Daniel, "Who Is Afraid of Russia Today?," December 5, 2014. As of April 25, 2016:
https://www.opendemocracy.net/od-russia/daniel-kennedy/who's-afraid-of-russia-today

Kevin, Deirdre, Francesca Pellicanò, and Agnes Schneeberger, "Television News Channels in Europe," European Audiovisual Observatory, October, 2013.

"KGB Practices. Folder 70. The Chekist Anthology," June 2007, History and Public Policy Program Digital Archive, Contributed to CWIHP by Vasili Mitrokhin. As of April 24, 2016:
http://digitalarchive.wilsoncenter.org/document/110319

Kivirähk, Juhan, "Integrating Estonia's Russian-Speaking Population: Findings of National Defense Opinion Surveys," International Centre for Defence and Security, December 2014.

Klishin, Ilya, "The Kremlin's Trolls Go West," *The Moscow Times,* May 21, 2014. As of April 25, 2016:
http://www.themoscowtimes.com/opinion/article/the-kremlins-trolls-go-west/500641.html

Knoema, "Leningrad Region—Gross Regional Product per Capita." As of April 24, 2016:
http://knoema.com/atlas/Russian-Federation/Leningrad-Region/GRP-per-capita

McDermott, Roger, “Russia Activates New Defense Management Center,” *Eurasia Daily Monitor,* Vol. 11, No. 196, November 4, 2014. As of April 24, 2016: http://www.jamestown.org/programs/edm/single/?tx_ttnews%5Btt_news%5D=43041&cHash=60f76b139c88bc8177cb5cba105c4fe6#.VrkNYhFUODE

Meek, James, “Putin’s Counter-Revolution,” *London Review of Books,* Vol. 36, No. 6, March 20, 2014. As of April 24, 2016: http://www.lrb.co.uk/v36/n06/james-meek/putins-counter-revolution

“Meeting of the Valdai International Discussion Club,” October 24, 2014. As of April 24, 2016: http://en.kremlin.ru/events/president/news/46860

“Military Doctrine of the Russian Federation,” 2014. As of January 19, 2017: https://www.offiziere.ch/wp-content/uploads-001/2015/08/Russia-s-2014-Military-Doctrine.pdf

Milne, Richard, “Party with Ties to Putin Pushes ahead in Estonian Polls,” *Financial Times*, February 27, 2015. As of April 24, 2016: http://www.ft.com/intl/cms/s/0/1decfbac-be8a-11e4-a341-00144feab7de.html#axzz3hEUPUbiQ

Ministry of Foreign Affairs of the Russian Federation, “Concept of the Foreign Policy of the Russian Federation,” February 12, 2013. As of April 24, 2016: http://archive.mid.ru//brp_4.nsf/0/76389FEC168189ED44257B2E0039B16D

“Moscow Bikers’ Club with Links to Putin May Expand to Estonia,” *bbn*, June 27, 2013. As of April 24, 2016: http://www.balticbusinessnews.com/article/2013/6/27/moscow-bikers-club-with-links-to-putin-may-expand-to-estonia

Myers, Steven Lee, “Russia Rebukes Estonia for Moving Soviet Statue,” *New York Times,* April 27, 2007. As of April 24, 2016: http://www.nytimes.com/2007/04/27/world/europe/27cnd-estonia.html?_r=0

NATO STRATCOM Center of Excellence, “Analysis of Russia’s Information Campaign against Ukraine,” Riga, 2015.

Nikolsky, Aleksey, “Russian ‘Spetsnaz’ Forces—from Saboteurs to Court Bailiffs,” *Moscow Defense Brief*, Centre for Analysis of Strategies and Technologies (CAST), No. 1, 2014.

“On the Border,” *The Economist*, March 7, 2015. As of April 24, 2016: http://www.economist.com/news/europe/21645845-how-nervousnessover-russia-affects-daily-life-and-politics-border

“On Human Rights. Folder 51. The Chekist Anthology,” June 2007, History and Public Policy Program Digital Archive, Contributed to CWIHP by Vasili Mitrokhin. As of April 24, 2016: http://digitalarchive.wilsoncenter.org/document/110120

"On the Russian Federation's National Security Strategy," December 31, 2015. As of January 19, 2017, http://www.ieee.es/Galerias/fichero/OtrasPublicaciones/Internacional/2016/Russian-National-Security-Strategy-31Dec2015.pdf

"Opros: Tret 'zhitelej Kryma ne podderzhivajut nezavisimost' Ukrainy" [Survey: 'Third of Crimeans Do Not Support Ukraine's Independence'], August 18, 2013. As of April 24, 2016: http://korrespondent.net/ukraine/politics/1593648-opros-tret-zhitelej-kryma-ne-podderzhivayut-nezavisimost-ukrainy

"O sisteme koordinacii, obespechivajushhej formirovanie edinogo informacionnogo prostranstva v sfere upravlenija informacionno-kommunikacionnymi tehnologijami v gosudarstvennom sektore, a takzhe izmenenii i priznanii utrativshimi silu nekotoryh aktov Pravitel'stva Rossijskoj Federacii" [On a System of Coordination to Ensure a Unified Information Space in the Area of Management of Information-Communication Technologies in the Government Sector, as Well as Amendments to Some Legal Acts of the Government of the Russian Federation], 2014.

"Party of Regions Hopes for Strengthening Collaboration with 'United Russia' Party," *KyivPost*, November 22, 2009. As of April 25, 2016: http://www.kyivpost.com/content/politics/party-of-regions-hopes-for-strengthening-collabora-53358.html

Pau, Aivar, "Estonia—Best Target for Cyber Crook," *Postimees,* June 15, 2015. As of April 24, 2016: http://news.postimees.ee/3226033/estonia-best-target-for-cyber-crook

Pavlovsky, Gleb, "Russian Politics Under Putin: The System Will Outlast the Master," *Foreign Affairs*, May–June 2016. As of October 4, 2016: https://www.foreignaffairs.com/articles/russia-fsu/2016-04-18/russian-politics-under-putin

Pelnens, Gatis, ed., "The 'Humanitarian Dimension' of Russian Foreign Policy Toward Georgia, Moldova, Ukraine, and the Baltic States," Riga: Centre for East European Policy Studies, International Centre for Defence Studies, Centre for Geopolitical Studies, School for Policy Analysis at the National University of Kyiv-Mohyla Academy, Foreign Policy Association of Moldova, International Centre for Geopolitical Studies, 2009, p. 18. As of July 5, 2016: https://www.academia.edu/1376506/The_Humanitarian_Dimension_of_Russian_Foreign_Policy_Toward._Georgia_Moldova_Ukraine_and_the_Baltic_States

Pifer, Steven, and Hannah Thoburn, "Nuanced Views in Eastern Ukraine," Brookings Institution, April 28, 2014. As of April 25, 2016: http://www.brookings.edu/blogs/up-front/posts/2014/04/28-nuanced-views-eastern-ukraine-pifer-thoburn

Podgornykh, Yuriy D., and Vasiliy Y. Dolgov, "Threats to Russia's Security: Controlled Chaos Technologies Are Coming to Replace Military Force," *Vozdushno-Kosmicheskaya Oborona Online,* October 13, 2014.

Podkopaev, Petr, Karen Dawisha, and James Nealy, Emglish translation of "The Reform of the Administration of the President of the Russian Federation" (originally published in Russian in *Kommersant*, May 5, 2000), n.d. As of March 8, 2018:
http://miamioh.edu/cas/_files/documents/havighurst/english-putin-reform-admin.pdf

Poleshchuk, Vadim, "Russian-Speaking Population of Estonia in 2014: Monitoring Report," Legal Information Centre for Human Rights, Tallinn, 2014.

Pomerantsev, Peter, *Nothing Is True and Everything Is Possible: The Surreal Heart of the New Russia*, New York: Public Affairs, 2015.

Pomerantsev, Peter, and Michael Weiss, "The Menace of Unreality: How the Kremlin Weaponizes Information, Culture and Money," The Interpreter Project of the Institute of Modern Russia, 2014.

Presentations at RAND Russian Measures Short of War Conference, February 2, 2016, Cambridge, UK.

Putin, Vladimir, "Russia and the Changing World," *Moskovskiye Novosti,* February 27, 2012. As of April 25, 2016:
http://www.mn.ru/politics/78738

"'Putin's Tycoon to Buy Media Empire in Serbia'—Daily," *B92*, February 3, 2016. As of April 24, 2016:
http://www.b92.net/eng/news/politics.php?yyyy=2016&mm=02&dd=03&nav_id=96924

Rapoza, Kenneth, "Here's What Putin's Counter-Sanctions Did to EU Exporters," *Forbes*, April 17, 2015.

———, "How Lithuania Is Kicking Russia to the Curb," *Forbes,* October 18, 2015. As of June 2, 2016:
http://www.forbes.com/sites/kenrapoza/2015/10/18/how-lithuania-is-kicking-russia-to-the-curb/4/#178da5ee251d

Ria Novosti, "Istorija agentstva," [RIA News, "Agency History"], n.d. As of April 24, 2016:
http://ria.ru/docs/about/ria_history.html#ixzz3miMy1Gng

Rossotrudnichestvo's homepage, n/d. As of April 24, 2016:
http://rs.gov.ru/en

"Rossotrudnichestvo razrabotalo doktrinu 'mjagkoj sily'" [Rossotrudnichestvo Developed the "Soft Power" Doctrine], June 10, 2014. As of April 25, 2016:
http://vz.ru/news/2014/7/10/694911.html

Roth, Andrew, "Former Russian Rebels Trade War in Ukraine for Posh Life in Moscow," *Washington Post*, September 16, 2015. As of April 25, 2016: https://www.washingtonpost.com/world/former-russian-rebels-trade-war-in-ukraine-for-posh-life-in-moscow/2015/09/13/6b71f862-3b8c-11e5-b34f-4e0a1e3a3bf9_story.html

Rozhkov, Evgeny, "[illegible]" [Russian Response to Estonia], Vesti.ru, January 17, 2017. Available at http://www.vesti.ru/doc.html?id=114941&cid=5

RT.com, "About Us." As of April 24, 2016: https://www.rt.com/about-us/

———. "Facts vs Fiction: Setting the Story Straight about RT." Last updated November 1, 2016: https://www.rt.com/facts-vs-fiction/#fvf-item-24

———, "More Money, More Problems?" February 19, 2015. As of April 25, 2016: https://www.rt.com/op-edge/233147-media-rt-financing-myths-facts/

———, "RT's 'Occupy' Coverage Nominated for Emmy News Award," August 22, 2012. As of April 24, 2016: https://www.rt.com/about-us/press-releases/rt-aib-awards-finals/; https://www.rt.com/news/rt-nomination-emmy-news-276/ https://www.rt.com/facts-vs-fiction/

"RT Is Very Upset with John Kerry for Blasting Them as Putin's 'Propaganda Bullhorn,'" April 25, 2014. As of April 24, 2016: http://www.businessinsider.com/john-kerry-rt-propaganda-bullhorn-russia-today-2014-4

"Russia Accused of Unleashing Cyber War to Disable Estonia," *The Guardian,* May 17, 2007. As of April 24, 2016: http://www.theguardian.com/world/2007/may/17/topstories3.russia

"Russian Bikers Help Estonia to Send Humanitarian Aid to Donbass." As of April 24, 2016: http://www.bbn.ee/?PublicationId=7601a8ee-e8d3-42c2-aa2e-4efb4ff60e4a

Russian Foreign Ministry, 2013.

"Russian Foreign Policy Concept," 2013. As of April 25, 2016: http://archive.mid.ru//brp_4.nsf/0/76389FEC168189ED44257B2E0039B16D

"Russian Military Doctrine 2014," December 25, 2014.

"Russians in Baltic States (2011)," by Xil, Licensed under CC BY-SA 3.0 via Commons. As of April 24, 2016: https://commons.wikimedia.org/wiki/File:Russians_in_Baltic_States_(2011).svg#/media/File:Russians_in_Baltic_States_(2011).svg"Russkaja Shkola Estonii" [Russian School of Estonia]. As of April 24, 2016: http://www.venekool.eu

Saivetz, Carol, "The Ties that Bind? Russia's Evolving Relations with Its Neighbors," *Communist and Post-Communist Studies,* Vol. 45, 2012, p. 402.

Schaefer, Olya, "Estonian Transit Shows Growth but Fears Russia," *The Baltic Times,* March 9, 2011. As of April 24, 2016: https://www.baltictimes.com/news/articles/28207/

Schenkkan, Nate, "Eurasian Disunion: Why the Union Might Not Survive 2015," *Foreign Affairs,* December 26, 2014.

Schmid, Fidelius, and Andreas Ulrich, "Betrayer and Betrayed: New Documents Reveal Truth on NATO's 'Most Damaging' Spy," *Spiegel Online,* May 6, 2010. As of April 24, 2016: http://www.spiegel.de/international/europe/betrayer-and-betrayed-new-documents-reveal-truth-on-nato-s-most-damaging-spy-a-693315.html

Sergunin, Alexander, and Leonid Karabeshkin, "Understanding Russia's Soft Power Strategy," *Politics,* Vol. 35, No. 3–4, 2015.

Shuster, Simon, "The Global News Network RT Is the Russian Government's Main Weapon in an Intensifying Information War with the West—and Its Top Editor Has a Direct Phone Line to the Kremlin," March 5, 2015. As of April 24, 2016: http://time.com/rt-putin/

———, "Inside Putin's On-Air Machine," *Time,* March 16, 2015, pp. 44–51.

Simon, Frédéric, "EU Plans Major Offensive to Diversify Gas Supplies," EurActive.com, June 2, 2015. As of October 4, 2016: https://www.euractiv.com/section/trade-society/news/eu-plans-major-offensive-to-diversify-gas-supplies/

Smith, David J., " 'Woe from Stones': Commemoration, Identity Politics and Estonia's 'War of Monuments,' " *Journal of Baltic Studies,* Vol. 39, No. 4, 2008.

Socor, Vladimir, "Alien Vandalism in Estonia's Capital," *Eurasia Daily Monitor,* Vol. 4, No. 85, May 1, 2007. As of April 24, 2016: http://www.jamestown.org/single/?tx_ttnews%5Btt_news%5D=32712&no_cache=1#.VuhvrVJUPJA

Soldatov, Andrei, and Irina Borogan, "Russia's Very Secret Service," *World Policy Institute,* Spring 2011. As of April 24, 2016: http://www.worldpolicy.org/journal/spring2011/russia-very-secret-services

Spriņģe, Inga, Donata Motuzaite, and Gunita Gailāne, "Spreading Democracy in Latvia, Kremlin Style," *Re:Baltica,* March 19, 2012. As of April 24, 2016: http://www.rebaltica.lv/en/investigations/money_from_russia/a/606/spreading_democracy_in_latvia_kremlin_style.html

Spruds, Andis, ed., "The Economic Presence of Russia and Belarus in the Baltic States: Risks and Opportunities," Riga: Centre for East European Policy Studies, Latvian Institute of International Affairs, Centre for Geopolitical Studies, Academic Center for Baltic and Russian Studies, 2012. As of July 5, 2016:
http://liia.lv/site/docs/The_Economic_Presence_for_WEB_atverums_2.1.pdf

Sputnik International, "About Us." As of April 24, 2016:
http://sputniknews.com/docs/about/index.html#ixzz3miRrfNB4

———, "Estonian Court Acquits Defendants in Bronze Soldier Protests," January 5, 2009. As of April 25, 2016:
http://sputniknews.com/world/20090105/119376653.html

"Ssora s Rossiej obernulas' dlja Estonii ogromnymi poterjami" [Fight with Russia Brought Estonia Great Losses], November 17, 2007. As of April 24, 2016:
http://zagolovki.ru/daytheme/estonia/17Nov2007

Standish, Reid, "Captured Russian Special Forces Soldier Describes His Unit Fighting in Eastern Ukraine," *Foreign Policy,* May 18, 2015. As of April 25, 2016:
http://foreignpolicy.com/2015/05/18/captured-russian-special-forces-soldier-describes-his-unit-fighting-in-eastern-ukraine/

Starikov Nikolaj, "Pochemu estoncy vedut sebja tak naglo, a Rossija tak sderzhanno?" [Why do Estonians Behave so Boldly and Russia so Timidly?], n.d. As of April 24, 2016:
http://www.contrtv.ru/print/2284/

Statistical Office of Estonia, Central Statistical Bureau of Latvia and Statistics Lithuania, "2011 Population and Housing Censuses in Estonia, Latvia, and Lithuania," 2015. As of April 24, 2016:
http://www.stat.ee/dokumendid/220923

Stokes, Bruce, "Russia, Putin Held in Low Regard Around the World," August 5, 2015. As of April 24, 2016:
http://www.pewglobal.org/2015/08/05/russia-putin-held-in-low-regard-around-the-world/

"Summit of Failure: How the EU Lost Russia over Ukraine," *Spiegel International,* October 16, 2013.

"Supplier Countries," 2016. As of October 4, 2016:
https://ec.europa.eu/energy/en/topics/imports-and-secure-supplies/supplier-countries

Sutyagin, Igor, "Russian Forces in Ukraine," RUSI Briefing paper, March 2015. As of April 24, 2016:
https://www.rusi.org/downloads/assets/201503_BP_Russian_Forces_in_Ukraine_FINAL.pdf

Tagel, Liisa, "Savisaar's Earlier Scandals," *Postimees,* September 23, 2015. As of April 24, 2016:
http://news.postimees.ee/3338007/savisaar-s-earlier-scandals

"Tallin. Itogi nochnyh besporjadkov" [Tallinn. Results of the Night's Disorder], April 27, 2007. As of April 25, 2016:
http://www.vesti.ru/doc.html?id=118604&tid=40252

"Tallinn: A Year Without the Bronze Soldier," *Kommersant*, April 25, 2008. As of April 24, 2016:
http://www.kommersant.com/p886482/r_1/foreign_relations/

Tamkin, Emily, "Russia's Senseless Investigation into Baltic Independence," *OpenDemocracy*, July 13, 2015. As of April 25, 2016:
https://www.opendemocracy.net/emily-tamkin/russia's-senseless-investigation-into-baltic-independence

Tashlykov, Sergei L., "General and Particular Features of Present-Day Conflicts Involving the U.S. and Its Allies," *Military Thought* 19, no. 3 (2010), pp. 60–69.

Tétrault-Farber, Gabrielle, "Looking West, Russia Beefs Up Spending on Global Media Giants," September 23, 2014. As of April 25, 2016:
http://www.themoscowtimes.com/news/article/looking-west-russia-beefs-up-spending-on-global-media-giants/507692.html

Thomas, Timothy, "Russia's Information Warfare Strategy: Can the Nation Cope in Future Conflicts?" *The Journal of Slavic Military Studies*, Vol. 27, No. 1, 2014, pp. 101–30.

Tikk, Eneken, Kadri Kaska, and Liis Vihul, "International Cyber Incidents: Legal Considerations," 2010 Cooperative Cyber Defence Centre of Excellence (CCD COE), 2010, pp. 71–74.

Trenin, Dmitri, "Russia's Breakout from the Post-Cold War System: The Drivers of Putin's Course," Carnegie Moscow Center, 2014. As of April 24, 2016:
http://carnegieendowment.org/files/CP_Trenin_Putin2014_web_Eng.pdf

Tsulaia, Jambul, "Real Talk with Russia's Fake Super Soldier," *The Daily Beast,* February 29, 2015. As of April 24, 2016:
http://www.thedailybeast.com/articles/2015/03/29/real-talk-with-russia-s-fake-super-soldier.html

TV Rain, Hard Day's Night, "Margarita Simonyan: 'Bolshe vsego menya besit vraniyo'" [Margarita Simonyan: "More Than Anything, I Get Outraged by Lies"], September 29, 2015. As of April 24, 2016:
https://tvrain.ru/teleshow/harddaysnight/intervju_s_glavnym_redaktorom_rt_margaritoj_simonjan-395356/

2011 Census in Estonia. As of April 24, 2016:
http://pub.stat.ee/px-web.2001/Dialog/statfile1.asp

"Ukraine: From Propaganda to Reality," Timothy Snyder, Chicago Humanities Festival, November 14, 2014. As of April 24, 2016: https://www.youtube.com/watch?v=eKFObB6_naw

"Union of Non-Citizens of Estonia." As of April 24, 2016: http://aliens.ee/?lang=en

U.S. Army Special Operations Command, n.d., pp. 15–17.

USASOC, "Counter-Unconventional Warfare: White Paper," September 26, 2014.

Van Herpen, Marcel, *Putin's Propaganda Machine: Soft Power and Russian Foreign Policy,* Lanham, Md.: Rowman & Littlefield, 2016.

Varfolomeyev, Oleg, "Ukraine Cuts Russian Gas Imports to a Minimum," *Eurasia Daily Monitor,* Vol. 13, No. 29, February 11, 2016. As of October 4, 2016: https://jamestown.org/program/ukraine-cuts-russian-gas-imports-to-a-minimum/

Vesti.ru, "Kiselev: cel' novoj struktury—vosstanovit' spravedlivoe otnoshenie k Rossii," [News.ru] [Vesti.ru, "Kiselev: The Goal of the New Organization—Is to Reinstate a Fair Image of Russia], December 9, 2013. As of April 24, 2016: http://www.vesti.ru/doc.html?id=1165191

"The Viktor and Vladimir Show," *The Economist,* February 14, 2015. As of April 24, 2016: http://www.economist.com/news/europe/21643194-germany-and-other-european-countries-worry-about-hungarys-ties-russia-viktor-and-vladimir

"Voennaja doktrina Rossijskoj Federacii" [Military Doctrine of Russian Federation]. As of April 24, 2016: http://rg.ru/2014/12/30/doktrina-dok.html

Volchek, Dmitry, and Daisy Sindelar, "One Professional Russian Troll Tells All," March 25, 2015. As of April 25, 2016: http://www.rferl.org/content/how-to-guide-russian-trolling-trolls/26919999.html

Walker, Shaun, "Russia's pro-Putin Biker Gang to Recreate WW2 in Epic Crimean Stunt Show," *The Guardian*, August 20, 2015. As of April 24, 2016: http://www.theguardian.com/world/2015/aug/20/russian-biker-gang-recreates-second-world-war-as-epic-stunt-show-crimea

Weaver, Courtney, "Malofeev: The Russian Billionaire Linking Moscow to the Rebels," *Financial Times,* July 24, 2014. As of April 24, 2016: http://www.ft.com/intl/cms/s/0/84481538-1103-11e4-94f3-00144feabdc0.html#axzz3qkBxTpj4

Website of President of Russia, "Visit to 'Russia Today' Television Channel," June 11, 2013. As of April 24, 2016: http://en.kremlin.ru/events/president/news/18319

"Why Russia's Soft Power Is Too Soft," *Russia in Global Affairs*, n.d. As of April 24, 2016: http://eng.globalaffairs.ru/redcol/Why-Russias-Soft-Power-Is-Too-Soft-15845

Wilson, Andrew, "'Political Technology': Why It Is Alive and Flourishing in the Former USSR," *openDemocracy*, June 17, 2011. As of April 25, 2016: https://www.opendemocracy.net/od-russia/andrew-wilson/political-technology-why-is-it-alive-and-flourishing-in-former-ussr

———, *Virtual Politics: Faking Democracy in the Post-Soviet World*, New Haven, Conn.: Yale University Press, 2005.

Winnerstig, Mike, ed., "Tools of Destabilization: Russian Soft Power and Nonmilitary Influence in the Baltic States," Sweden Defense Research Agency (FOI), December 2014. As of April 25, 2016: http://www.foi.se/rapport?rNo=FOI-R—3990—SEF

World Bank, "GINI Index (World Bank Estimate)." As of April 24, 2016: http://data.worldbank.org/indicator/SI.POV.GINI

Yu, Filimonov G., "Aktual'nye voprosy formirovaniya strategii 'myagkoi sily' vo vneshnei politike Rossiiskoi Federatsii" [Current Issues in the Formation of "Soft Power" Strategy in the Foreign Policy of the Russian Federation], *Zakon i pravo*, No. 9, 2013, pp. 9–18.

Zdanavičius, Liudas, and Matthew Czekaj, eds., *Russia's Zapad 2013 Military Exercise: Lessons for Baltic Regional Security*, Washington, D.C.: Jamestown Foundation, 2015, pp. 9, 14–19, 96.

Zhukov, Yuri, "Rust Belt Rising: The Economics Behind Eastern Ukraine's Upheaval," *Foreign Affairs*, June 11, 2014. As of April 24, 2016: https://www.foreignaffairs.com/articles/eastern-europe-caucasus/2014-06-11/rust-belt-rising

CHAPTER FOUR

Case Study: Iran

Political warfare is a central component of the Islamic Republic of Iran's strategy to resist and defeat U.S. influence in the Middle East. Iran's perceived military disadvantages relative to the U.S. strongly influences its use of political warfare, which includes information operations, use of media, political manipulation, and economic domination. According to Iranian officials, the Islamic Republic's ideology of resistance and its skill in conducting political warfare allow it to offset U.S. power in the Middle East. In particular, they believe that Iran's position as a "natural" power in the region and leadership of an "axis of resistance"—composed of Lebanese Hezbollah, the Syrian regime, pro-Iranian Palestinian groups, and several Shi'a Iraqi militias—allows it to counter American "imperialism" in the region.

Amid the transformations taking place in the Middle East in recent years, Iran has employed a robust political warfare strategy in an effort to expand its regional power. The 2003 overthrow of Saddam Hussein and the installment of Shi'a Iraqi parties in Baghdad provided Iran with unprecedented influence in Iraq. The rise of the Islamic State in Syria and the Levant has also provided Tehran with unique opportunities to wield political power in its western neighbor. In defending Syria's Bashar al-Assad government, Iran has also engaged in political warfare, albeit in a more challenging context that lacks a significant Shi'a population and sees the Damascus regime besieged by foreign-backed militants. Nevertheless, Iran's political warfare skills will continue to present a unique challenge to U.S. interests for the foreseeable future. Iran's ideology of resistance may have lost much of its appeal since the Arab uprisings and the Syrian war, as Iran's reputation in the

Arab world has declined markedly since those events started. But Iran continues to champion marginalized Shi'a groups under the threat of Sunni jihadism. This alone provides Tehran with substantial influence and leeway in the Arab world.

This chapter will first explore the philosophy underpinning Iran's use of political warfare. The chapter will then examine the means by which the Iranian regime employs its strategy in the political, religious, cultural, economic, and military spheres, with a particular emphasis on Iranian activities in Iraq and Syria.

Iran's Views of Political Warfare

Iran's historical memory of foreign manipulation, combined with its sense of national importance, continues to shape its conduct of political warfare today. The Iranian regime believes itself to be under a permanent state of siege, while at the same time fulfilling its role as a defender of the downtrodden in the face of Western imperialism.

A one-time superpower, Persia had become a relatively weak and backward nation-state by the nineteenth and twentieth centuries. It was never colonized by Western powers, but British, Russian, and then American interference in its internal affairs was a rather common and, from an Iranian viewpoint, humiliating occurrence. Iran's 1979 revolution, which overthrew the authoritarian U.S.-backed Shah Mohammad Reza Pahlavi, was at its heart an anti-imperialist struggle.

Today, Iran's current supreme leader Ayatollah Ali Khamenei believes that the United States is opposed to the very "essence" of the Islamic Republic rather than simply its regional activities. In particular, Khamenei believes that Washington is seeking to foment a "velvet revolution" against the regime through the promotion of Western democracy, support for civil society, and psychological warfare.[1] In September 2015, Khamenei warned a group of Islamic Revolutionary

---

[1] Khamenei believes that the United States was behind the so-called color revolutions that occurred in several former Soviet republics, including Ukraine and Georgia (Khamenei.ir, 2016).

Guards Corps (IRGC) commanders: "Economic and security infiltration is not as important as intellectual, cultural, and political infiltration" ("Enemy's Infiltration," 2015). According to the supreme leader, one of the main techniques of the evil empire's soft war is to create discord among the people of a society.[2]

The Islamic Republic has adopted an anti-imperialist liberation ideology as a core pillar of its foreign policy. According to the Iranian elite, the United States has replaced Great Britain as the Middle East's chief imperial power. Having "liberated" Iran from Western imperialism, the Islamic Republic feels it must now help free its Muslim brethren across the region from the yoke of foreign control.

Iran's political warfare is aimed at countering the United States, Saudi Arabia, Israel, and ISIL at the same time; many in Iran view each to be part of a broad conspiracy to defeat the Islamic revolution. According to Khamenei, the United States seeks to maintain instability in the Middle East in order to justify its continued presence (Khamenei, 2015). Iranian officials believe Israel to be at the forefront of Western colonialism in the Middle East. The Islamic Republic has long championed the Palestinian cause and made it a central part of its political warfare in the region. But as the region's conflicts have increasingly broken down along the Sunni-Shi'a divide, Iranian leaders have viewed Saudi Arabia as their chief ideological rival due to the kingdom's espousal of Wahhabism, an interpretation of Sunni Islam that considers the Shi'a to be heretics. The Iranians blame the Saudis for the rise of ISIL, whose ideology tracks closely with the Saudi official religion. Many of the Iranian elite also argue that ISIL was created, either indirectly or directly, by American actions in the region. Khamenei argues that the United States sees the spread of extremist groups such as ISIL as a means of breaking up countries into smaller, more easily controlled pieces (Khamenei, 2015). Therefore, the conflicts enveloping the region are not sectarian in nature, but "a war between those devoted to the goals of the United States and the West versus supporters of the independence of nations" ("Tahlil-e Rahbar-e," 2014). The Chair of the Supreme Council for

2 Khamenei's sensitivity to perceived Western political warfare greatly increased after the 2009 Green Movement protests (Gholamzadeh, 2015; Wastnidge, 2014).

National Security, Ali Shamkhani, claims that the United States supports ISIL in order to "instill fear and instability" among the people of the region and to prevent them from "determining their country's destiny" for themselves ("Firoozabadi," 2014). And Armed Forces Chief of Staff Hassan Firoozabadi has questioned why it was so "easy" for ISIL to conquer Mosul, claiming that the capture of the key Iraqi city was planned by the United States and Israel in cooperation with former Baathist generals (Fars News Agency, 2014).

For Iran, having soft power is more important than having hard power. Shaping people's morals and desires is a more durable form of power than simple diplomatic treaties. According to Khamenei, while countries most often seek support from others through economic, political, and military means, that is only a source of temporary influence (Haghighatkha, 2013). Once the treaties and contracts end, the influence ends. Rather, leveraging shared culture, religion, and civilization is longer lasting. According to Khamenei, "The principal target of the Islamic Republic are the peoples" of the region (Haghighatkha, 2013). Influencing hearts and minds can speed up and deepen the strategic influence of Iran, and it is less costly. Thus, among the panoply of political warfare tactics, the subset of "soft power" tactics aimed at cultural and religious influence represent the heart of the Iranian approach.

But in reality, Iran's conduct of political warfare is also motivated by its inherent weakness in the face of U.S. power, although it is never depicted as such by Iranian officials. Iran is a developing country with limited economic and military capabilities, especially when compared to the United States and its regional allies. Therefore, it must revert to asymmetric tactics. In addition to supporting proxy groups such as Hezbollah and Iraqi Shi'a militias, Iran views political warfare to be an integral nonmilitary asymmetric capability that can offset American military, economic, and political influence.

Iranian officials are sanguine about Iran's strength in the region, which they deem to be a result of the success of its ideology. According to the narrative, Iran's soft power—specifically regarding the religious and cultural realm—is increasing to the point that it could become "a consequential cultural axis in the region and the world" (Hossein and Alamdari, 2008). In Khamenei's words, "This strength, this [regional] influence, is precisely because of the revolution. Were it not for the revolution; were it not for the revolutionary spirit; were it not for the revolutionary actions, this influence would not exist" (Khamenei, 2015). Khamenei argues that if foreign powers are removed from the region, the people—whose will is influenced by Iran's Islamic revolution—will inevitably come to power (Abdollahian, 2013).

Methods and Agencies of Political Warfare

Over the past several decades, the Islamic Republic has developed a range of political warfare tactics that it employs to counter U.S. influence, the most predominant of which are using proxy militias to manipulate local politics to Tehran's advantage; employing religious ideology to recruit and inspire militants; using economic influence as a means of political leverage; engaging in psychological warfare and propagating the Islamic revolution's ideology through its national broadcasting network, NGOs, and loyal proxies; engaging in cultural and religious diplomacy efforts; and co-opting grassroots movements. The extent to which Iran emphasizes each tactic depends on the political and social context of each target country.

Iran's proxy militias play a prominent role in Tehran's political warfare strategy in ways that complement their asymmetric military capabilities. Iran's expeditionary force, the Islamic Revolutionary Guards Corps Qods Force (IRGC-QF), leads these efforts. It is responsible for training pro-Iranian militants across the Middle East and beyond. Through loyal militias, Iran intimidates political actors and manufactures political crises that it must then be called in to resolve. It has encouraged militias to establish political organizations, with Lebanese Hezbollah serving as the most successful example of this tactic. At the same time, Iran backs and often establishes competing militias in order to provide it with strategic options and to prevent any one group from becoming too powerful. A constant proliferation of new militias may also be intended to confuse the enemy as to their origins, thus providing Iran with plausible deniability for violent actions. Ambiguity regarding the level of Iranian control over the groups could also serve to make local actors' Islamic resistance ideology appear to be merely inspired by the Islamic Republic rather than Tehran-directed propaganda. Meanwhile, in their training camps, the IRGC-QF attempts to impart to trainees the spiritual inspiration its fighters experienced on the front lines of the Iran-Iraq War.

Iran also engages in political warfare through economic means, focusing especially on gaining market share in weak economies and using its newfound influence to pressure host governments. For instance, when Baghdad cracked down on pro-Iran militias in 2008, Tehran exploited Iraq's dependence on Iranian electricity by cutting off power to Basra (Sullivan, 2009). In Syria, Iran has taken advantage of Assad's isolation by stipulating that the billions of dollars in credit extended to Damascus be used mostly on purchasing Iranian goods and services (Yazigi, 2015). However, Tehran rarely exerts economic pressure in such overt ways, making it difficult to assess the extent to which economic power is being leveraged for political warfare purposes. But at the same time, the Islamic Republic's control over so much of the Iranian economy—either through state-owned enterprises, IRGC-affiliated companies, or multibillion-dollar religious foundations (*bonyads*)—provides it with opportunities to incorporate its business ventures within an overall political warfare strategy.[3] Often, Iran appears to have used enticements rather than threats to achieve its political objectives. Throughout the presidency of Mahmoud Ahmadinejad, Tehran sought to escape international isolation by providing economic incentives to African and Latin American countries, likely in exchange for diplomatic support of its nuclear aspirations and anti-imperialist stance. For instance, beginning in the mid-2000s, Iranian economic cooperation and investment in Mauritania increased in parallel with the African country's diplomatic move away from Israel (Szrom, 2010). Meanwhile, Tehran rewarded Bolivian president Evo

---

[3] Religious foundations are not subject to public financial scrutiny, and are answerable only to Supreme Leader Khamenei. The Foundation of the Oppressed and War Veterans owns hundreds of subsidiaries, through which it has invested in energy, engineering, and multiple other industries throughout the world (Thaler et al., 2010; Ilyas, 2009).

Morales's anti-American stance with promises of a $1 billion investment in its weak industrial sector (Garcia and Hudson, 2007).

Psychological and ideological operations are integral to Iran's foreign policy. As a government borne out of revolution, it understands the importance of propaganda (Eistenstadt, 2010). According to Michael Eisenstadt, "Whereas the United States undertakes information and psychological operations to support its military operations, Iran frequently undertakes military activities (i.e., displays of force and surrogate terrorist operations) to support its propaganda and psychological warfare operations" (Eistenstadt, 2010, p. 2). As such, it most often seeks victory through demoralizing its enemy rather than through traditional military victory—for instance, conducting a war of attrition against American troops in Iraq and Afghanistan or terrorizing enemy civilians. Iran's depiction of the 2006 Lebanon war as a victory for Hezbollah is an example of its information operations.

The Islamic Republic of Iran Broadcasting (IRIB) is a central component of Iran's information operations. IRIB controls all Iranian broadcasting, as the constitution prohibits private broadcasting networks (Pahlavi, 2012, pp. 21–33). IRIB has branches in 45 countries, and it has five foreign language news channels (Pahlavi, 2012, pp. 1–14). It also provides some programming for Hezbollah's Al-Manar station (Wastnidge, 2014, pp. 1–14). Iran's use of IRIB to target regional audiences became more pronounced in 2003, following the U.S. invasion of Iraq, when IRIB established the 24-hour Arabic-language channel Al-Alam. The channel has over 50 reporters in around 40 countries, with offices in Gaza and Ramallah. It broadcasted around-the-clock coverage of the 2006 Lebanon war in an attempt to enhance the "axis of resistance's" influence throughout the Middle East.[4] Iran launched its English-language channel PressTV in 2007. PressTV is dedicated to improving Iran's international image and countering American "propaganda" across the world. The channel receives over $25 million per year from IRIB (Pahlavi, 2012, pp. 21–33). In 2011, IRIB launched its Spanish-language channel HispanTV, which airs throughout Latin

---

[4] IRIB's other Arabic-language channel is Al-Kowthar, which was created in 2006 and focuses on Shi'a religious programming (Harsin, Tavisarkani, and Jafari, 2009, pp. 225–69).

America. Then-IRIB head Ezatollah Zarqami claimed that HispanTV would play "a major role in reflecting the ideological legitimacy of our system to the world" (Black, 2010). Iran tailors its channels to the tastes of the intended audience. For instance, while PressTV and HispanTV feature female anchors with loose headscarves for their Western audiences, Al-Alam anchors wear a conservative style of *hejab* common in the Arab world ("Presidente Italiano," 2016; Pahlavi, 2012, pp. 21–33).

Iran also engages in cultural and religious diplomacy efforts. Its diplomacy during the first decade of the Islamic Republic was marked primarily by efforts to export its religious ideology through various institutions, including the Islamic Propagation Organization. However, by the 1990s, and following the Iran-Iraq War, Iran realized that Arab publics were more interested in Arab unity than Islamic unity. Therefore, the "drive for disseminating revolutionary propaganda transformed over time into a more subtle and organized effort to introduce Iran's culture and values abroad" (Maltzahn, 2013, p. 210). In an attempt to coordinate its cultural diplomacy efforts, Iran created the Islamic Culture and Relations Organization (ICRO) in 1995. The organization's mission is to introduce the Islamic revolution's values and realize "Islamic unity by strengthening cultural relations among Muslim states" (Maltzahn, 2013, p. 205). According to an article in the Iranian journal *Political Science*, the ICRO cultural attaché has an "important duty to [build up] public opinion and [forge relationships with] social elites and transmit Islamic-national values" in Middle Eastern countries (Harsin, Tavisarkani, and Jafari, 2009, p. 239). At the same time, the position can also provide effective cover for intelligence operatives attempting to recruit from among local populations. For instance, Iran's former cultural attaché in Argentina, Mohsen Rabbani, is the leading suspect behind the 1994 bombing of a Jewish community center in Buenos Aires ("Iran Set Up Terrorist Networks in Latin America," 2013). Khamenei, through his direct control of the IRIB and the ICRO, exercises strong influence over Iran's soft power activities.[5]

[5] The Supreme Leader appoints the head of IRIB, who also serves on the governing council of ICRO. Moreover, the ICC staffs are subordinate to the Supreme Leader's representatives in each country (Wastnidge, 2014, pp. 1–14; Maltzahn, 2013).

Prominent in Tehran's cultural diplomacy efforts are the approximately 70 Iranian Cultural Centers (ICCs) around the world that are run by the ICRO but often operate out of Iran's embassies (Maltzahn, 2013). In most countries, the ICCs either run or help facilitate Persian-language training programs. They also host Iranian cultural events for the public. Each center's ideological activities are tailored to each country's circumstances (Maltzahn, 2013). In Shi'a-populated countries, the centers' ideological focus tends to be on Shi'a solidarity, while in non-Shi'a Muslim countries, the centers focus instead on pan-Islamic issues. For instance, because of the lack of a sizeable Shi'a population in Syria, the ICCs there have historically focused on Persian language and culture as well as promoting Islamic unity. On the other hand, the ICC in Beirut, which is home to many Shi'a, has long held Ashura processions, commemorating the martyrdom of the Shi'a Imam Hussein.[6] Meanwhile, in non-Muslim countries, such as those in Latin America and Africa, ICCs focus on general anti-imperialist and anti-Zionist propaganda ("Tahghighaat-e Diin," 2010). In Sierra Leone, for example, the ICC hosts an annual Qods Day celebration to denounce Israel ("International Quds Day," 2013).

In countries with an established Shi'a population, Tehran also recruits clerics to act as ambassadors of Islamic Republic ideology. Iran funds their training in the seminaries of Qom before returning them home to serve as religious leaders in their communities. The Ahl al-Bayt Assemblies (ABA), which are also run by the ICRO and are located throughout the world, are involved in these activities. Iran's efforts to expand its influence over Shi'a communities have reached as far as Latin America—especially to Argentina and Brazil, which are home to sizeable Lebanese Shi'a diaspora populations. For instance, the Brazilian city of Foz do Iguaçu is home to the Ayatollah Khomeini Shrine, which is overseen by the Qom-affiliated cleric Sheikh Muhammad Khalil (Comemoração do Eid al Fiter em Foz do Iguaçu [Commemoration of Eid al-Fitr in Foz do Iguaçu], 2014).

---

6 Interestingly, the ICC in Lebanon is not as active as the Damascus one, possibly due to the fact that there is already a history of people-to-people exchange between Lebanese and Iranian Shi'a that does not require government assistance to flourish (Maltzahn, 2013).

Despite the coordinated efforts by ICRO and IRIB, several Iranian analysts have highlighted what they view to be weaknesses in Iran's cultural diplomacy. One criticism is that Iran's efforts are still dominated by official statements (Golshan-Pezhou and Reza, 2012). Some analysts also claim that Iranian programming lacks creativity—a significant problem in the face of what they see as the West's patient and well-organized efforts to blacken the image of the Islamic Republic (Golshan-Pezhou, 2012). According to Mahmoud Reza Golshan-Pezhou, a foreign policy analyst at the Expediency Council's Center for Strategic Research, Western media has been successful in making people frightened of Iranians by focusing on human rights abuses and labeling Iran a destabilizing force (Golshan-Pezhou, 2012). Golshan-Pezhou recommends that Iran look to examples from other countries that have been successful in exporting their culture. For instance, China has created over 300 Confucius Institutes around the world, which promote Chinese culture and language. It has also invested heavily in improving its English-language state media, committing to increase the number of its English-language television employees tenfold by 2016 and increasing the number of its stations, especially in the United States and Africa. Golshan-Pezhou also highlights Turkish soap operas and Bollywood films as invaluable means of gaining soft power for Turkey and India, respectively. With patience and correct management, he believes such types of cultural diplomacy will allow the Islamic Republic to expand its soft power regionally and internationally.

A new initiative undertaken by IRIB may be intended to address these weaknesses. In 2014, Iran began production on an Arabic-language film titled *Songs of My Homeland*, which attempts to counter ISIL's propaganda (Scotten, 2015). The film's plot—a love story about a Christian woman and a Muslim man living under the threat of ISIL—as well as its Arabic-language dialogue suggest that the project intends to convey to Arab audiences that Iran is concerned with the welfare of all the region's inhabitants, and not just Shi'as. It also is a way to redirect Arab concerns about Iran onto the West; the film's director, Abbas Rafei, said he made sure to "show ISIL's American weapons and to say that most of their equipment was provided by the Zionists" ("Hadaf az Saakht-e," 2015).

Amid the recent transformations taking place in the Arab world, the Islamic Republic has attempted to take credit for the grassroots mobilizations. According to Iranian officials, examples of Iran's success abound. The Arab Spring, for instance, was depicted as part of an "Islamic Awakening" inspired by Iran's revolution. According to Deputy Foreign Minister for Arab Affairs Hossein Amir Abdollahian, leaders of the Egyptian revolution were inspired by the Islamic revolution and the role Iran currently plays in the region (Abdollahian, 2013). Abdollahian asserts that, with access to more media, Arab youth were able to see how Iran has progressed, how it stands up to the West to safeguard its indigenous scientific achievements. They saw how Iran is an example of a government governed by the will of the people. And in September 2014, following the fall of the Yemeni capital of San'aa to Houthi rebels, prominent Iranian parliamentarian Ali Reza Zakani boasted that "three Arab capitals have today ended up in the hands of Iran and belong to the Islamic Iranian revolution," and he expressed optimism that San'aa would become the fourth ("Sanaa Is the Fourth," 2014). Zakani apparently was referring to Damascus, Baghdad, and Beirut, the latter of which is heavily under Hezbollah influence.

Iran's Political Warfare in Iraq

Iran's political warfare activities in Iraq following the U.S. overthrow of Saddam Hussein were driven by the concern that a strong, democratic Iraq allied with Washington, coupled with a revived center of Shi'ism in Najaf, would challenge the Islamic Republic's survival and its attempts to become a paramount regional power. To prevent a new Iraq from posing a threat to Iran, Tehran required an Iraq that was weak and under Iranian, rather than American, influence.

In the face of overwhelming U.S. military power, Iran turned to an asymmetrical military strategy coupled with extensive political warfare, where it felt it had a relative advantage. Led by IRGC-QF commander Qassem Soleimani, Tehran has pursued a multipronged political warfare strategy: It has cultivated relationships with a variety of Shi'a and non-Shi'a political parties, exerted influence over their decisionmaking, and played various factions against one another, thus making itself indispensable as an arbiter of political gridlock.[7] Tehran has also used its substantial influence in the Iraqi economy to buy off political factions as well as counter Iraqi policies that run contrary to Iran's interests. It has established loyal militias through which it can violently pressure politicians while simultaneously cultivating some of them to attain political office themselves. Finally, through its militias and loyal junior clerics, Iran has sought to increase its ideological and religious influence in Iraq to counter Najaf's Grand Ayatollah Ali al-Sistani and his "quietist" school of Iraqi Shi'ism, which shuns direct clerical participation in politics. Sistani's stance is a rebuke of Islamic Republic founder Ayatollah Khomeini's doctrine of *velayat-e faqih*, which calls for clerics to oversee both the spiritual and political affairs of the nation.

Despite Iran's successes in waging political warfare, attempts to dominate Iraq have nonetheless run up against Iraqi nationalist sentiment among the Shi'a and a lack of general enthusiasm for *velayat-e faqih*. At the same time, the fall of Mosul to ISIL in June 2014 forced Baghdad to rely on Iran for protection, a turn of events that has presented the Islamic Republic with an even greater opportunity to put its political warfare strategy into practice.

Political Manipulation

Iran has strived to become the ultimate arbitrator of Iraqi politics, capable of both creating crises and intervening to resolve them (Felter and Fishman, 2008, p. 32). Tehran has adopted a strategy of flexibility by cultivating several political groups and militias, only to encourage loyal factions to split off from the original organization once it has attempted to moderate its position or move away from Iran. Once a new militia has become established, Iran has sought to place its members within the government as an added source of influence. It has also shown an interest in having pro-Iran militias manipulate protest movements and steer them in a direction favorable to Tehran.

---

[7] Leveraging his connections with political actors across the ethnic and religious spectrum, "Soleimani has been directly involved in nearly all major Iraqi government deliberations since the fall of Saddam" (Nader, 2015, p. 5).

Following the fall of the Baathist regime in Baghdad, Iran sought to empower the former opposition groups—including the Shi'a groups Dawa and the Supreme Council for the Islamic Revolution in Iraq (SCIRI)—that it had once harbored from Saddam.[8] Iran, however, does not support only Shi'a parties but has also built strong ties to Kurdish leaders in northern Iraq, such as former Iraqi president Jalal Talabani and his Patriotic Union of Kurdistan (PUK), whom Tehran aided in their struggle against Saddam (Brennan et al., 2013). For over a decade, Tehran has largely succeeded in corralling its allies to run on united lists, thus forming dominant blocs in parliament (Nader, 2015). One such list was the United Iraqi Alliance (UIA)—dominated by SCIRI and Dawa—which won the majority of seats in the 2005 interim parliament tasked with drafting Iraq's constitution. Thus, Iran was able to ensure that the constitution was favorable to its interests.[9] Following the 2010 parliamentary election, Qassem Soleimani helped Nouri al-Maliki keep his position as prime minister by convincing Ibrahim al-Jaafari's Iraqi National Alliance to join Maliki in opposing Ayad Allawi, despite the latter's Al-Iraqiyya bloc having won the most seats.[10]

Iran's relationship with Moqtada al-Sadr serves as another example of its flexibility in working with a wide variety of actors—all with the objective of strengthening Tehran's influence in Iraqi affairs and positioning itself as the indispensable resolver of political disputes. The son and nephew of two late grand ayatollahs, Moqtada al-Sadr has a strong popular base that, in addition to the fact that he spent the Saddam

---

8 At the beginning of the post-Saddam era, SCIRI was the closest group to Iran. SCIRI's former armed wing, the Badr Corps, had fought alongside the IRGC during the Iran-Iraq War. Dawa also has strong connections with Iran because some of its leaders, including former Prime Minister Nouri al-Maliki, spent time there during their exile. However, the party's leadership is split over how close Dawa should be to Iran (Vaezi, 2008).

9 The Iraqi constitution allows provinces to call for a referendum on autonomy, allowing southern Shi'a provinces to fall closer into Tehran's orbit (Nader, 2015).

10 Another means by which Iran gained influence was through the Accountability and Justice Committee, which, at least during Maliki's administration, was led by politicians close to Tehran. In a role similar to that of Iran's Guardian Council, the committee has disqualified many Sunni candidates with real or suspected Baathist pasts (Nader, 2015).

years in Iraq rather than in Iran, means that he has not felt it necessary to align with the Islamic Republic as closely as some of his Shi'a rivals (Vaezi, 2008). He in fact portrays himself as a fierce Iraqi nationalist and has criticized Iran's interference in Iraqi politics (Eisenstadt, Knights, and Ali, 2011). Nonetheless, Iran played an instrumental role in establishing Sadr's Jaish al-Mahdi (JAM) militia in 2004 (Stanford University, n.d.). Moreover, in 2008, Qassem Soleimani brokered a ceasefire between JAM and the Maliki government, and Iran allowed Sadr to move to Qom, where he studied in the seminary until 2011 (Allam, Landay, and Strobel, 2008).

Meanwhile, Iran remained supportive of Maliki's rule, despite his increased sectarian rule, which is widely blamed for playing a role in the ISIL takeover of much of the country's Sunni-inhabited areas.[11] Today, despite Maliki's failures, and despite having dispatched Iran's Supreme National Security Council chair Ali Shamkhani to Najaf to help facilitate a peaceful transition of Iraqi power to Haider al-Abadi, Khamenei has not abandoned the former prime minister (Shabani, 2015). This is because Maliki still holds sway within the Dawa party leadership and among several powerful pro-Iran Shi'a militias such as Asa'ib Ahl al-Haq (AAH), and thus he is able to act as a spoiler to thwart policies counter to Iranian interests. Prime Minister al-Abadi, on the other hand, wants to maintain a balance of relations between Iran and the United States (Hiltermann, 2015; ""Sistani Questions Qassem Soleimani's Influence," 2015). Maliki's importance to Iran was illustrated further in the fierce opposition from pro-Iran groups to Abadi's August 2015 announcement of reforms, one of which would eliminate the deputy prime minister post that Maliki currently holds (Malas, 2015). Abadi also called for an end to the quota system that reserves the post of prime minister for the Shi'a. AAH chief Qais al-Khazali responded by calling for an adaptation of the presidential system—one that likely would bring Maliki back to power (Habib, 2015). Moreover, shortly after Abadi's announcement, Iran indicated its strong support for Maliki by inviting him to meet with Khamenei in Tehran ("Sistani

---

[11] Maliki's sectarian policies included purges of Sunni military officers and violent attacks on Sunni protesters (Solomon and Nissenbaum, 2014).

Questions Qassem Soleimani's Influence," 2015). In the face of heavy pressure from Iran's allies in Iraq, Abadi has been unable to follow through on his promised reforms. In addition to preserving Maliki's power, this also has benefited Iran by disillusioning Abadi's supporters and making him appear weak (Malas, 2015).

Despite these successes, Iran has had to contend repeatedly with Iraqi nationalism, even among the Shi'a. In 2007, deciding that a close association with Iran was preventing the organization from achieving broad appeal, SCIRI's leadership changed the party's name to the Islamic Supreme Council in Iraq (ISCI) and replaced Iran's Supreme Leader Ali Khamenei with Sistani as its spiritual guide (Felter and Fishman, 2008). However, in 2012, and in a move that is typical of Iran-backed organizations, the Badr Organization faction within ISCI split off as its own party—one devoted to Iran's Supreme Leader (Habib, 2012). Meanwhile, playing on Iraqi nationalist sentiment has given Moqtada al-Sadr the political capital to remain independent of Iran.

Economic Influence

Since the fall of Saddam, Iraq's economy has become more reliant on Iran, providing Tehran with additional political leverage. Iraqi dependence on Iran for cooking gas, heating oil, gasoline, and electricity has made it particularly vulnerable to pressure from Tehran (Knights, 2010). In the spring of 2008, for instance, Iran cut Basra's electricity to retaliate against Baghdad's crackdown on pro-Iranian militias there (Sullivan, 2009).

Moreover, in a mutually reinforcing way, Iran's successful political warfare strategy has fed its overall economic power in Iraq. In an op-ed in *Fars News*, Hossein Tavassoli notes that Iran's influence over Iraqi politics has led to an increase in bilateral trade (Tavasoli, 2015). Iraq is now Iran's second-largest trading partner, and relies on the Islamic Republic for most of its cement, tiles, ceramics, dairy products, and electricity (Bozorgmehr, 2015). Iran's Economics Minister predicts that bilateral trade will increase to $20 billion in "the near future" (including transit goods and tourism) (Bozorgmehr, 2015). To help this along, Iran plans by 2017 to create a 37,400-hectare free trade zone near the border with Iraq, primarily for oil-related industries, steel factories, power plants, and renewable energy (Bozorgmehr, 2015).

Support for Iraqi Militias

Armed Iraqi Shi'a militias are key instruments of Iran's political warfare strategy. Iraqi Shi'a militias are often more dependent on Tehran than on Iraqi Shi'a political parties, providing Iran the opportunity to instill its revolutionary ideology and bypass the traditional centers of religious and political power in Iraq. Iran's support for the militias originally was borne out of its desire to ensure that the Iraqi government would not become too powerful or independent. But Ayatollah Sistani's June 2014 *fatwa* calling on all Iraqi men to join the fight against ISIL has also created a new pool of volunteer fighters that Iran can cultivate to counter the Iraqi clerical establishment.

Since the end of the Saddam era, Tehran has used militias in a variety of ways to increase its political power over Baghdad. The evolution of these militias has tended to follow a pattern by which a group emerges to pressure politicians through violence before ultimately establishing a political wing of its own that competes for political office. Once a group becomes politically entrenched to the point that it either is too powerful for Iran to control or seeks to temper its violent activities, Tehran often encourages factions to splinter off from the party to establish new, more subservient militant organizations (Eisenstadt, Knights, and Ali, 2011). The rise of multiple Shi'a militias also benefits Iran as it creates ambiguity regarding which groups are under their control. Not only does this provide plausible deniability for violent actions, but it could also allow the militias to pursue policies that further Iran's interests without provoking an Arab nationalist backlash (Levitt and Smyth, 2015).

Today, around 50 Shi'a militias operate in Iraq (Nader, 2015). Some of these are openly loyal to the Islamic Republic, while others' links to Iran are less clear. Since 2003, the IRGC-QF and Lebanese Hezbollah have provided training in Iran for thousands of Iraqi militants.[12] Religious and ideological indoctrination are a mandatory part of the courses, although the trainees apparently take these lessons with varying levels of seriousness (Felter and Fishman, 2008). It also appears that many of the militants prefer the Hezbollah trainers in Iran because they speak Arabic and do not display the Persian chauvinism of the IRGC trainers.

The most influential of the Iraqi Shi'a militias is the Badr Organization, an outgrowth of SCIRI's Badr Corps, which fought alongside Iran during the Iran-Iraq War. The militia's current leader, Hadi al-Amiri, served with Qassem Soleimani on the front lines (Brennan et al., 2013). The group is fiercely loyal to Supreme Leader Khamenei and, since 2004, has been operating Islamic Cultural Centers in Baghdad and other Iraqi cities to propagate the Islamic Republic's ideology (Nader, 2015). By 2007, and in an illustration of Iran's strategy to infiltrate the Iraqi government with loyal militia members, the Badr Organization had gained control of the Ministry of Interior's border control and major crimes units, and was vying for control of the ministry's intelligence office (Parker, 2007). From 2010 to 2014, al-Amiri served as Maliki's Minister of Transportation, a post from which he facilitated Iranian flights carrying weaponry to Syria (Fulton, Holliday, and Wyer, 2013). In 2014, Badr Organization member Mohammed al-Ghabban was appointed Interior Minister (Parker, 2015).

Because Moqtada al-Sadr's grassroots popularity has allowed him a great deal of independence from Iran, despite having benefited from IRGC-QF and Hezbollah arms and training to establish his JAM militia in 2004, Iran has cultivated splinter groups comprising former Sadrist militia members.[13] One such group is Asa'ib Ahl al-Haq (AAH), formed in 2006 and led by Sadr rival Qais al-Khazali (Brennan et al., 2013). AAH has proven to be a valuable political warfare tool for Iran,

[12] The courses range from 20-day basic paramilitary training to advanced paramilitary skills for those advancing to leadership positions (Felter and Fishman, 2008).

[13] While studying in Qom, Sadr disbanded much of JAM, creating a social services organization, the Muhamidoon, in its place. However, following the fall of Mosul to ISIL, Sadr established the Peace Brigades. This new militia is funded directly by Sadr and does not appear to receive assistance from Iran (Brennan et al., 2013; Stanford University, n.d.; Habib, "Senior Iraq Cleric 'Saves' the Government").

performing violence on behalf of Tehran, winning seats in the Iraqi parliament, and spreading the Islamic Republic's ideology in Iraq's seminaries. The militia, which aims to advance Iran's Islamic revolution within Iraq, came to be known as one of several "special groups" that targeted U.S. forces in Iraq. The U.S. government considers AAH to be a "direct-action" arm of IRGC-QF (Brennan et al., 2013, p. 125). By 2011, the group was receiving upward of $5 million per month in cash and weapons from Iran (Habib, "Senior Iraq Cleric 'Saves' the Government"). As of 2014, there were between 1,000 and 5,000 AAH militiamen (Morris, 2014).

In December 2011, al-Khazali announced AAH's decision to participate in Iraq's political system (Wyer, 2012). AAH won seats in the parliament in 2014 as the Al-Sadiqun Bloc, and Maliki included it in his State of Law Coalition (Nader, 2015). AAH has also sought to manipulate popular grievances and public protests in order to further Iran's agenda. Two days after the outbreak of anti-corruption demonstrations in Iraq at the end of July 2015, for instance, Khazali expressed his support for the protesters and announced the establishment of AAH civilian units to help facilitate reform, prompting some civil society activists to express concern that the militia sought to hijack the protest movement (Habib, 2015). The next month, following Abadi's announcement of reforms, which AAH opposed, the militia's leaders instructed members to participate in demonstrations without wearing their uniforms, making them appear as popular demonstrators (Habib, 2015). Some militias also put up posters in Baghdad's Tahrir square accusing the original protesters of being former Baath party members who opposed Islam and were trying to return to power (Habib, 2015).

Kata'ib Hezbollah (KH) is another JAM splinter group established by the IRGC-QF in 2007 (Nader, 2015). KH leader Jamal al-Ibrahimi (aka Abu Mahdi al-Muhandis) is a close advisor to Soleimani. As of 2014, KH had 3,000 fighters, all of whom swore an oath of loyalty to Khamenei. KH are elite fighters who eschew involvement in overt politics and social services. In addition to targeting U.S. forces, Iran has also used KH to fight Iranian opposition groups in Iraq.[14]

Tehran does not exert control over all of the Iraqi Shi'a fighters, however. The competition for influence over the approximately 100,000 Iraqi Shi'a recruits—known as Popular Mobilization Units (PMU), or *Hashd Shaabi*—should be seen as part of the broader struggle between Ayatollahs Sistani and Khamenei (Parker, 2015). Despite claims of unity, the PMU fighters are split into a variety of factions.

One group of PMU militias is referred to as *Hashd Sistani* due to their loyalty to Ayatollah Sistani (Hiltermann, 2015). These fighters are paid by Sistani and are not seen to have overt political ambitions (Habib, 2015). Although they are not considered to be as well trained as the Iran-backed units, they comprise around half of all PMU fighters. The three major *Hashd Sistani* militias were created by the chiefs of the Iraqi Shi'a shrines.[15]

The Iran-backed portion of the PMU consists of two broad groupings referred to as *Hashd Soleimani* and the *Third-Term Hashd*—the latter because of its support for a Maliki third term as prime minister (Hiltermann, 2015). These militias receive the best equipment and salaries out of all the PMU fighters (Habib, 2015). While AAH and Badr Organization spokesmen claim that all PMU fighters are under Iraqi government control and that arms are provided to them only by the Abadi government, Badr's control over the Ministry of Interior likely provides the pro-Iran groups with a conduit to their superior weaponry. One Interior Ministry official estimates that 70 percent of the ministry employees are loyal to militias, many of which are likely Iran-affiliated groups (Parker, 2015).

Further illustrating Iranian control over its PMU fighters, one Iraqi Ministry of Defense official has complained that the Iran-affiliated militias "refuse to cooperate with the army when it comes to preparing plans and deployment" (Habib, 2015). According to U.S. military offi-

---

14 The Iranian hardline website Ya-Lesarat al-Hossein claims that KH carried out the September 2013 attack against the Mujahedin-e Khalq's Camp Ashraf in Iraq, which killed around 50 members belonging to the Iranian dissident group (Gholipour, 2014).

15 The Imam Hussein Shrine in Karbala has fielded the Ali al-Akbar Brigade. Karbala's Abbas Shrine created the Abbas Battalion, and the Imam Ali Shrine in Najaf has established its own militia (Steele, 2015).

cials, even the Fifth Iraqi Army Division is under an Iran-backed militia chain of command, and rarely communicates with the Defense Ministry (Parker, 2015). Teenagers being trained in the Iran-affiliated militia camps are being indoctrinated by clerics promoting the Islamic Republic's ideology. The training has been likened to Iran's ideological indoctrination of its people's militia, the Basij Resistance Force (Qaidaari, 2015). ISCI and Moqtada al-Sadr also control parts of the PMU.

Both the Sistani-aligned militias and Iran-backed groups have cultivated Sunni units. This is likely out of an understanding that ISIL can be defeated only if Sunni grievances are taken into account.[16] On the Iranian side, the Badr Organization, KH, and AAH all have Sunni fighters (Al-Salhy, 2015). These groups have deployed Sunni battalions of 250 to 600 fighters in Diyala, Anbar, and Salah al-Din provinces (Al-Salhy, 2015). Sunni fighters (most likely those affiliated with the well-equipped pro-Iran PMUs) have lauded the arms and training they have received from the Shi'a militias compared to the long waiting period for Ministry of Defense assistance (Al-Salhy, 2015). According to one Sunni commander, not teaming up with a Shi'a militia means "[n]o weapons, no equipment and no political future" (Al-Salhy, 2015). Iran likely will attempt to corral these Sunnis into a political bloc in the future. However, Tehran does not want to see the Sunnis become too powerful; its militia and Iraqi government allies have pushed back, for example, against Ayatollah Sistani's support for the government's proposal to create a National Guard, which would put Sunni militias under provincial control (Habib, 2015).

Religious and Ideological Propaganda

Since the fall of the Baath regime in Baghdad, Tehran has sought to increase its economic and religious clout in Iraq's holy cities as a means of furthering the Islamic Republic's ideology. This was in response to the expected post-Saddam resurgence of the Najaf school of quietist Shi'ism, which Tehran viewed as a major challenge.

---

[16] The pro-Sistani Ali Akbar Brigade claims that 16 percent of its fighters are Sunni, while its allied Abbas Battalion is thought to be 5 percent Sunni (Steele, 2015).

By 2003, Sistani had become the most revered Shi'a cleric in the world, far surpassing Ayatollah Khamenei in religious clout. Since Saddam's removal, Sistani's influence has only grown further with the unfettered interactions between Iranian and Iraqi Shi'a.[17] Moreover, as the security situation in Iraq has deteriorated, Sistani has found it necessary at times to put aside his quietist tendencies and intervene in the nation's political affairs. He has come out in strong support of Prime Minister Abadi's reform proposals, which could counter Iranian influence in Iraq by bringing an end to the sectarian quota system that has favored Shi'a politicians close to Tehran (Hiltermann, 2015). In fact, Sistani and his fellow Najaf clerics reportedly regret their part in bringing Shi'a political parties to power following the fall of Saddam, having witnessed the role that Maliki's sectarian governance played in driving Sunni grievances (Hiltermann, 2015; Arango, 2013). Sistani is also reported to have written a letter to Khamenei criticizing Qassem Soleimani for intervening in Iraqi political affairs (Hiltermann, 2015).

Despite Tehran's challenges on the religious front, the Iranians have been successful in taking over a substantial share of the Najaf and Karbala economies, which has provided it with yet another avenue through which to pursue its political agenda. In 2005, Iraq signed an agreement allowing 5,000 Iranian pilgrims into the country each day.[18] The political influence resulting from such large-scale pilgrimages is not lost on Iranian officials. A December 2015 article published in the reformist Ebtekar newspaper noted that the numerous Iranian officials accompanying Shi'a pilgrims in a Karbala Ashura procession—all protected from ISIL by Iran-backed militias—was an effective form of "Iranian-style diplomacy" (Roozbahani, 2015). In the past decade, Iranian religious foundations and government-affiliated firms have invested heavily in Najaf and Karbala; as of 2007, Iran was providing $20 million per year to improve Najaf's tourism infrastructure

---

[17] By 2006, an estimated 80 percent of Shi'a around the world were seeking spiritual guidance from Sistani, bringing the cleric over $700 million per year in religious taxes. This estimate comes from clerical offices in Qom as well as clerics with access to pilgrimage polling numbers (Khalaji, 2006; Slackman, 2006).

[18] An average of 40,000 Iranian pilgrims visit Iraq each month, with 3 to 4 million coming during the holy month of Moharram (Dagher, 2009; Eisenstadt, Knights, and Ali, 2011).

(Sullivan, 2009). Moreover, Iran has benefited from its influence over the Iraqi government to obtain lucrative contracts in the holy cities. For instance, the Iranian government-owned company Setad Bazsazi Atabat Aliyat has been responsible for shrine renovations and the construction of a hospital, multiple hotels, and an Iranian seminary in Karbala (Dagher, 2009). The Iranian firm Shamsa has been said to hold a "virtual monopoly" over the multi-million-dollar Iranian pilgrim industry (Dagher, 2009). The local partners whom Shamsa chooses for the pilgrims' transportation, protection, and accommodation are mostly affiliated with pro-Iran Shi'a political parties, effectively creating a revolving door that empowers parties loyal to Tehran.

Iran is also preparing for the death of the 85-year-old Sistani. It appears to be grooming Ayatollah Hashemi Shahroudi to be among his top successors in Najaf (Nader, 2015). The Iraqi-born Shahroudi is an original leader of SCIRI and a former chief of Iran's judiciary. He is very loyal to Ayatollah Khamenei. In October 2011, Shahroudi opened an office in Najaf, marking an intensification of Iran's attempts to influence the holy city's seminary students. Shahroudi is reportedly trying to attract Sistani's students by paying higher stipends than the senior cleric, and his followers are constructing "what could be Najaf's largest seminary" (Al-Kifaee, 2012). However, Shahroudi will not likely be able to gain Sistani's clout due to his espousal of the Islamic Republic's *velayat-e faqih* spiritual-political doctrine, which does not resonate among the majority of the world's Shi'a (Nasr, 2007).

Therefore, Iran is also cultivating junior Iraqi clerics in a potential attempt to "fragment the Iraqi Shi'ite religious network and force Najaf to become politically and economically subordinate to Qom and Iranian clerical authority" (Wyer, 2012, p. 22). One such minor cleric is Seyyed Mohammad Tabatabai, the deputy head of AAH, who propagates Iranian ideology in Iraqi seminaries and is critical of Sistani (Nader, 2015). AAH also operates a network of religious schools called "Seal of the Apostles" and has established the Department of Religious Schools in Najaf, which recruits young clerics (Nader, 2015). Sheikh Akram al-Kaabi is another junior cleric loyal to Ayatollah Khamenei (Hashem, 2015). As the head of Hezbollah al-Nujaba, a militia that splintered off from AAH, he appears in social media in both clerical garb and military fatigues.[19]

Iran also is planning on opening Iranian Cultural Centers in Karbala and Najaf, which are likely to propagate Iran's ideology alongside Persian language and culture programs ("Dafaatar-e Rayzani-ye," 2015). At the same time, Tehran is turning the Islamic Republic founder Ayatollah Khomeini's former Najaf residence into a museum ("Ailing Ayatollah," 2015).

In addition to their attempts to influence the Iraqi religious sphere, the Iranian government and its Iraqi allies engage in propaganda efforts to spread their "resistance" ideology. IRIB's Arabic news channel Al-Alam—accessible by antenna to all Iraqi households—serves as Iran's primary means of spreading propaganda in Iraq (Pahlavi, 2012).

Through their various media outlets, Tehran and pro-Iran politicians and militias continue to accuse Washington of backing ISIL as an excuse to bring U.S. forces back to Iraq. Shi'a politicians in parliament are fanning this conspiracy theory through speeches and social media in a rhetorical campaign that Colonel Steve Warren, U.S. military spokesman in Baghdad, refers to as "part of the Iranian propaganda machine" (Sly, 2015). In February 2015, KH's Al-Etejah Press website ran a story claiming that U.S. forces were air-dropping supplies to ISIL ("Babil Council Member," 2015). It is, in fact, now common among both Shi'a and Sunni Iraqis to believe that the United States is supplying ISIL with food and weaponry (Sly, 2015). Whether this is the result of Iran's propaganda efforts or whether Iran's rhetoric simply is attempting to tap into suspicions already in place is unclear. Regardless, these beliefs place Prime Minister Abadi in a difficult situation when accepting U.S. help.

Tehran and its allies have also used their media outlets to depict the Islamic Republic as Iraq's most dependable ally. The narrative was helped by the fact that, while Tehran sent weapons and IRGC advisors to the beleaguered Baghdad government only 48 hours after the fall of Mosul, the U.S. air strikes did not begin until two months later (Daragahi et al., 2015). As the IRGC-QF took command of much of

---

19 See Al-Nujaba's Twitter feed. As of January 19, 2017: https://twitter.com/alnujabaiq.

the fighting, even the traditionally secretive Qassem Soleimani became a viral sensation on Iranian and pro-Iranian militias' social media (Lamothe, 2015).

The Iraqi Nationalist Challenge

Regardless of the level of political influence Tehran wields over Baghdad, Iran will always have to contend with Iraqi nationalism. In 2009, for instance, Karbala residents gathered in the streets to protest the award of a multimillion-dollar contract to the Iranian company Al Kawthar to renovate the city's historic center. A Shi'a hotel owner interviewed said, "We are Arabs, we will not accept to be colonized by anyone."[20]

A 2010 poll found that only 18 percent of Iraqi Shi'as held a positive view of Iran's ties with Iraq's political leaders (Nader, 2015). For their part, Iranian analysts are aware of the challenge posed by Iraqi nationalism. In a 2008 article, Mahmoud Vaezi, Deputy for Foreign Policy Research at the Expediency Council's Center for Strategic Research (and current Minister of Information and Communications Technology), noted that only in times of sectarian tension could a shared Shi'a identity paper over Iraqi nationalism.[21]

In the end, the Islamic Republic's ultimate challenge may be to maintain enough sectarian tension in Iraq that the Shi'a will be forced into Tehran's arms, while preventing massive instability that could spill over inside Iran's borders.

---

[20] That same year, Iraq's Interior Minister banned the use of foreign language signs in Najaf and Karbala, which was a response to concerns over the prevalence of Persian signs throughout the holy cities (Dagher, 2009).

[21] Iraq affairs analyst Cyrus Barna Boldaji also made this point in an article for *Iranian Diplomacy*. According to Boldaji, in an Iraq where all factions, including Sunnis, participate in elections, one can predict that such a plurality of parties could be elected to office that they would lessen Iran's ability to influence Baghdad. Boldaji warns that one should also prepare for the possibility that technocrats and secular politicians could emerge who would play up nationalism in order to attract votes. He concludes, "One can predict that if, in the midterm, a new Iraq is to go to war with a country, that country could once again be Iran" (Vaezi, 2008; Boldaji, 2010).

Iran's Political Warfare in Syria

Syria serves as an integral part of the "resistance axis" in that it provides Iran with access to Lebanese Hezbollah's stronghold in southern Lebanon. Losing Syria would hinder Iran's ability to provide Hezbollah with weaponry, significantly undermining the Islamic Republic's ability to deter an Israeli or U.S. attack. Moreover, the loss of Iran's only Middle Eastern state ally would be a great blow to Iranian prestige and power throughout the Arab world.[22] Therefore, Tehran prefers to keep the Damascus regime intact. However, if Assad were to fall, Iran would need loyal militias to help maintain control over at least the western portions of Syria that are contiguous with Hezbollah territory—a potentiality that Iran's political warfare strategy likely is intended to address. Iran's assistance to Syria comes in addition to what the Assad regime receives from Hezbollah, which has deployed around 6,000 fighters, especially to the areas around Damascus, Homs, Aleppo, and Hama (Kajjo, 2015). However, as the war has dragged on, and Hezbollah has lost upward of 1,000 fighters, it has faced trouble persuading its members to remain in Syria (Rosenfeld, 2015).

Tehran has engaged in political warfare in Syria as part of its efforts to ensure the survival of a pro-Iranian government in Damascus. To buttress its position, Iran has employed the following tactics: attempting to indoctrinate the National Defense Force (NDF) militias with Islamic revolutionary ideology, appealing to foreign Shi'a fighters' desire to protect Syria's holy shrines, taking advantage of Syria's economic dependency to increase Tehran's influence over the Damascus government, and engaging in public diplomacy to endear Syrians to the Islamic Republic. However, Iran's successes in the political and ideological realm have been limited by the relative lack of advantages as compared to Iraq. Iran and Syria do not share a common border that has facilitated centuries of cultural exchange. Nor does Syria have a sizeable Shi'a population that could provide Iran with substantial inroads into its society. Moreover, Bashar al-Assad's Baathist regime and his

22 The Baghdad government's interest in maintaining balanced relations between the United States and Iran means that Tehran cannot count on Iraq as a true ally.

popular base are predominantly secular and unlikely to be swayed by the Islamic Republic's appeals to pan-Islamic unity. And while Tehran benefited from the influx of thousands of Shi'a recruits heeding Ayatollah Ali al-Sistani's call to arms against ISIL in Iraq, Sistani and the clerical establishment in Najaf have issued no such edict regarding Syria. Also dampening support for Assad among the general Iraqi Shi'a populace is the Syrian leader's support for Sunni fighters during the Iraqi sectarian civil war following Saddam's ouster (Ghazi and Arango, 2012).

Training the National Defense Force

The IRGC-QF has played an essential role in shaping Iran's overall strategy in Syria, including its political warfare strategy. Led by General Soleimani, an estimated 2,000 IRGC officers have been dispatched to serve mostly as advisors to the Syrian armed forces (Daragahi, "Iran's Intervention in Syria"). Most importantly, the Qods Force was responsible for establishing and training the NDF, the network of Syrian people's militias that serve as the backbone of Assad's defense against the rebels. Established in late 2012, the NDF grew out of Assad's desire to formalize and professionalize the amalgam of local militias called the Popular Committees.[23]

According to IRGC commander Mohammad Ali Jafari, the late Brigadier General Hossein Hamedani was instrumental in helping the Syrians organize around 100,000 of these fighters (*IR Diplomacy*, 2015). Because NDF units are comprised of locals from the areas from where they are based, each unit reflects the religious makeup of its locale, meaning that the NDF is made up of Alawites, Christians, Druze, Armenians, Shi'a, and even Sunnis (Pyruz, 2016). As a result, while Iran may be able to cultivate ties with some of the units it trains, it is unlikely that the NDF as a whole would emerge as an Iranian proxy should Assad's regime fall.

Nevertheless, Iranian officials tend to exaggerate the ideological influence the Islamic Republic has over Syrian forces. They attempt to

---

[23] Syria sends many of the NDF fighters to Iran for two-week urban warfare training. The fighters are flown from Latakia air base to Tehran, and then bussed to a secret location (Fulton, Holliday, and Wyer, 2013; Solomon, 2013).

portray Assad's successes as the result of Iran's transfer of its Islamic resistance culture to the Syrians. For instance, according to Hamedani, it was Iran that convinced Assad to establish the Popular Committees that eventually became the NDF ("Aakharin Mosahebe-ye," 2015). This is despite the fact that the Popular Committees already had a precedent in Syrian history; according to Middle East analyst Chris Zambelis, Bashar al-Assad's "reliance on irregular armed citizen militias . . . mirrors his father's strategy of employing irregular citizen militias as auxiliaries to the regular Syrian army and security forces in his fight against the Syrian Muslim Brotherhood" (Zambelis, 2015). Moreover, although the NDF is largely secular, Iranian officials often liken it to Iran's religiously conservative people's militia, the Basij Resistance Force. In Hamedani's telling, the Syrians were originally skeptical of the benefits of a volunteer force, but after seeing their battlefield prowess, they eventually came to appreciate the Basiji spirit.[24]

Despite these claims, Iran's political warfare efforts do not appear to have been successful in mobilizing a substantial number of Syrian fighters who are loyal to the Islamic Republic. Iran's limited religious and ideological influence in Syria means that Tehran would have to rely to a large extent on foreign Shi'a fighters to further its interests if Assad were to fall.

Religious Appeals to Shi'a Fighters

Central to Iran's political warfare in Syria have been efforts to portray the fight as essential to the defense of the Syrian Shi'a and their shrines in Damascus—a strategy that contradicts the Islamic Republic's general preference to downplay the appearance of sectarian bias, and one that highlights Tehran's need to attract foreign fighters.

In the event that the Assad regime falls, Iran will have to rely on militias to help it maintain strategic territory in western Syria. While some NDF units trained by the IRGC-QF may come to Tehran's aid,

---

[24] Hamedani also claimed that another sign of Iran's success in Syria is the transfer of the spirit of the Sacred Defense—Iran's term for the Iran-Iraq War—to the front lines there. According to him, whereas the Syrian military had banned praying and Quran readings at the front, now the Syrian soldiers are praying and participating in religious ceremonies ("Aakharin Mosahebe-ye," 2015; "Tashkiil-e Goroohi," 2015).

the Iranian regime likely realizes that it would have to rely largely on the thousands of foreign fighters, from over a dozen pro-Iranian Shi'a militias, currently fighting in Syria (Chulov, 2014). Although these militias often appear to be operating independently, Phillip Smyth suggests that "they should be recognized as subnetworks of a broader IRGC-Qods Force network and part and parcel of a larger regional strategy" (Smyth, 2015, p. 56). Moreover, the pro-Iranian nature of the Shi'a fighters traveling to Syria is the result of the broader competition between Iran's Ayatollah Khamenei and the Iraqi Ayatollah Sistani; while Sistani called his followers to arms in the fight against ISIL in Iraq, he has refused to authorize their travel to Syria to defend a Baathist dictator allied with Tehran. For its part, Iran has pursued an extensive social media recruiting campaign justifying its involvement in Syria in terms of defending the country's minority Shi'a population and the revered Sayyeda Zainab Shrine in Damascus from destruction by Sunni extremists.

Around a third of the foreign Shi'a fighters in Syria come from Iraq, where Iran has been cultivating militias for over a decade (Daragahi, "Iran's Intervention in Syria"). The first wave of Iraqi fighters, in the summer of 2012, came from the well-established "special groups," including AAH and KH (Chulov, 2014). At that time, Iran also encouraged its Iraq-based militias to establish committees in southern Iraq and Diyala to recruit fighters for Syria (Ghazi and Arango, 2012). By the fall of 2012, thousands of Iraqi Shi'a militants were traveling to Iran and then being flown to Damascus (Ghazi and Arango, 2012). Others were taking pilgrimage tour buses from Iraq to Damascus. In spring 2013, Iran began its own concerted internet-based recruitment effort (Smyth, The Shiite Jihad in Syria and Its Regional Effects). Illustrating the routine way in which Iran creates new militias to fit new contexts, in the summer of 2013, KH established a Syria-based militia called Liwa Abu Fadl al-Abbas (LAFA) (Levitt and Smyth, 2015). This militia now serves as the major hub for a network of smaller Shi'a militias operating in Syria.

In early 2014, Iran showed its ability to adapt to a fluid battlefield environment in Syria when it substantially increased the deployment of Afghan Shi'a fighters to compensate for the loss of Iraqi Shi'a militants who had returned home to fight ISIL (Smyth, The Shiite Jihad in Syria and Its Regional Effects). Iran had established the Afghan Fatemioun militia in 2011, drawing on the approximately 2,000 Hazara Afghan refugees already living in Syria as well as the hundreds of thousands of refugees in Iran (Smyth, The Shiite Jihad in Syria and Its Regional Effects). Tehran has recruited most of the Fatemioun from areas around the Iranian cities of Mashhad and Qom, attracting some with monthly salaries and promises of residency permits (Dehghan, 2015). Many other Afghan refugees have been threatened with deportation if they do not fight in Syria, with some as young as 12 years old being forced to deploy.[25] As part of their training, clerics attempt to instill the importance of defending the Shi'a shrines, although Afghan fighters have been deployed throughout Syria, including near the Golan Heights. Today, the Fatemioun Division numbers around 12,000 fighters (Dehghan, 2015). As of May 2015, at least 200 Fatemioun fighters had been killed in Syria. Iran has also attracted up to 1,000 Pakistani Shi'as to Syria ostensibly in order to defend the Sayyeda Zeinab Shrine.[26]

Beginning in July 2015, there was a resurgence of Iran-backed Iraqi Shi'a militia recruitment efforts on social media, which may have been timed to coincide with increased Russian involvement in Syria; later that month, Qassem Soleimani was in Moscow meeting with Vladimir Putin to discuss Russia's future role in the war (Smyth, 2015; Parker, 2015). This new round of recruitment and deployment saw a leading role by new groups, likely because the more established Iraqi groups were focused on the fight in Iraq. For instance, KH splinter group Kataib al-Imam Ali (KIA) pursued an active social media campaign and placed billboards around Najaf calling for fighters in Syria (Levitt and Smyth, 2015). Hezbollah al-Nujaba, an outgrowth of AAH, was another Iraqi Shi'a militia involved in recruiting fighters for Syria (Smyth, "Iran-Backed Iraqi Militias Are Pouring into Syria").

In order to overcome Assad's lack of popularity among Iraqi Shi'a, Iran and its militias have pursued an overtly sectarian recruitment

---

25 The majority of Afghans appear to be trained at an IRGC base in Varamin, 60 kilometers south of Tehran ("Iran Sending Thousands," 2016).

26 The so-called Zeinabiyoun Pakistani militants were recruited through social media advertisements promising salaries of over $1,000 per month (Dehghanpisheh, 2015).

drive. In 2013, a pro-AAH and KH music video claimed, "Bashar is not our concern, it is the Shi'ites" (Smyth, The Shiite Jihad in Syria and Its Regional Effects, p. 8). Several clerics in Qom have issued fatwas urging the Shi'a to fight in Syria. In November 2013, the Iranian Ayatollah Kazem Haeri became one of the first Shi'a clerics to issue a public religious edict endorsing jihad in Syria ("Why Iraq's Shiites Are Taking Jihad to Syria and Supporting Al-Assad," 2013). Junior Iraqi clerics cultivated by Iran, such as AAH's Mohammad al-Tabatabai, also have endorsed the jihad and made their way to the Syrian front lines (Al-Jaffal, 2013). Meanwhile, neither Sistani nor the rest of the Najaf clerical establishment have authorized fighting in Syria (Mamouri, 2013).

Central to recruiting Shi'a fighters and justifying Iran's presence in Syria has been the Sayyeda Zeinab shrine in Damascus. The shrine is the tomb of Zeinab, the revered daughter of Imam Ali, and is thus a sacred site for all Shi'a.[27] The shrine's location is also strategically critical for Assad; according to a report from the Institute for the Study of War, "Without Sayyeda Zeinab, the opposition could form a contiguous area of control encircling regime positions in western Damascus and would cut off regime access to Damascus International Airport" (Fulton, Holliday, and Wyer, 2013, p. 25). Even when fighters are killed away from Damascus, their role defending the Zeinab shrine is highlighted ("Enemy's Infiltration," 2015).

Economic Influence

While the Syrian conflict has been a costly endeavor for Iran, which has spent billions of dollars defending the Damascus regime, Assad's reliance on Iran as Syria's economic lifeline has its advantages. The Syrian market provided an outlet for Iranian goods amid crushing international sanctions against Iran. More importantly, Iran's expanded market share is likely to grant Tehran political influence in Damascus to last long after the civil war.

Four years of international isolation has forced Syria to shift its economy fundamentally away from Turkey and Qatar, and realign

---

[27] Dating back to the 1980s, the Zeinab shrine has also served as a gathering place for Iran-backed Shi'a militants, serving as a transfer hub for Shi'a Saudis traveling to Lebanon and Iran for training (Smyth, The Shiite Jihad in Syria and Its Regional Effects).

squarely with Iran (Yazigi, 2015). As recently as 2010, bilateral trade between Iran and Syria stood at only $320 million, while Syria-Turkey trade had reached $2.5 billion (Yazigi, 2015). In 2011, shortly before the uprising, Syria rejected an application by the IRGC-affiliated Toseye Etemaad-e Mobin, which was competing with Turkish and Arab companies for a mobile phone license in Syria (Yazigi, 2015). The Iranian company was the only foreign firm rejected by Syria, suggesting that Assad may have been concerned with providing the IRGC an expanded role in his country.[28]

However, as relations became strained with Turkey and Qatar over their support for anti-Assad rebels, Syria's economy became captive to Iran. Since 2011, Iran has provided Assad with billions of dollars in loans and credit for oil and other imports (Black, 2015). These loans have come with strings attached. When in January 2013, Iran extended a $1 billion line of credit to Assad, Tehran stipulated that 60 percent of the imports paid for through the credit line had to come from Iran (Yazigi, 2015). Similarly, an August 2013 credit line of $3.6 billion was limited to oil products, most of which would come from Iran (Yazigi, 2015). Today, Syrian government contracts often are open only to Iranian companies (Yazigi, 2015). Many of these are lucrative contracts to rebuild the country's war-ravaged infrastructure. As a result, Iran-Syria trade reached $1 billion in 2014, with 35 percent of Syria's imports coming from Iran (Lucas, 2015; Black, 2015). In May 2015, Iran announced that it would be extending another $1 billion line of credit to Assad (Lucas, 2015). In addition to loans, Iran is now seeking collateral from the Syrian government in the form of real estate assets and state properties (Yazigi, 2015). Locals in Damascus also report that Iranians are buying up real estate in the city center at a rapid rate (Black, 2015).

---

28 Further illustrating the Damascus government's reticence toward Iranian economic influence, Syrian Foreign Minister Walid Muallem warned U.S. Secretary of State John Kerry in 2010, "If we don't succeed in opening up our economy, you'll come back here in ten years and you'll meet with *Mullah* Assad" (Remnick, 2015).

Public Diplomacy

Efforts to endear the Syrian people to the Islamic Republic form another component of Tehran's political warfare strategy. However, attempts to forge bonds between the Syrian and Iranian publics do not appear to have met with overwhelming success.

Concerted attempts to pursue cultural exchange are relatively new, having begun only after the Islamic revolution and mostly at the behest of the Iranians, who were determined to export their revolutionary ideals (Maltzahn, 2009, pp. 33–50). Tehran's outreach historically has focused on Iranian Cultural Center (ICC) programs that introduce Syrians to Persian language and literature as well as "Iranian Islamic culture" (Maltzahn, 2009, pp. 33–50). The programs also promote student and academic exchanges. The ICC in Damascus, which opened in 1983, is Iran's most active in the world (Maltzahn, "Iran's Cultural Diplomacy"). However, interest in the Damascus center's activities appears limited, with most of the students studying Persian at the ICC being minority Shi'a (Maltzahn, 2009, pp. 33–50). According to an ICRO study, the areas with a higher density of Alawites, such as the coastal region, are more open to Iran's efforts in the cultural realm (Maltzahn, "Iran's Cultural Diplomacy"). In 1996, Iran opened a second center in Latakia.

Charitable activities appear to form a core part of Iran's public diplomacy strategy in Syria. Tehran has assigned each Iranian province a Syrian province for which to gather donations of cash, food, and medical supplies ("Tashkil-e Setaad-e," 2014). For instance, the Iranian province of Hamedan has been tasked to care for the people of Hama province. As of October 2015, the people of East Azerbaijan province had provided around $1.5 million in cash and an anticipated $100,000 in food and noncash donations to the Syrian people ("Ersaal-e Komakha-ye," 2015). The donations are often delivered by religiously conservative Iranian nongovernmental organizations (NGOs) on the ground in Syria. For example, the Center for a Better Tomorrow (CBT), which was founded in 2013 by Iranian university and seminary students, has delivered aid to Syrians in Tartous, Latakia, Damascus, and Homs ("Komakha-ye Kheiri-ye," 2014). An October 2014 CBT ceremony held in Tartous illustrates the ideological messaging that can accompany Iranian charity. As the organization handed out food and blankets to 120 households—with each household receiving a package worth $200—the head of the CBT told the crowd, "As representatives of the revolutionary people of Iran, we support the Syrian people's resistance against the terrorists and mercenaries of the Zionists and the U.S. There is no doubt that the goals and path of [our] two nations are one."

The International Union of Unified Ummah is another Iranian organization founded by "university students" likely affiliated with the Basij ("Didaar-e Azaa-ye," 2014). The organization advocates pan-Islamic unity. The union has been providing food and medicine to Palestinians living in Syria's Yarmouk refugee camp ("Komakha-ye Kheirieh-ye Iran beh Soorieh Hargez Motevaghef Nashod," 2015). In April 2014, members of the group formed the Peace Pilgrims Caravan and visited Homs to deliver food and toys to the city's children ("Didaar-e Azaa-ye," 2014). The Damascus ICC is also seeking ways to help the Syrian people.

In March 2015, IRGC Brigadier General Hamedani also touted the establishment of the Jihad al-Banaa (Iranian officials refer to it as Syria's "Construction Basij"), which is tasked with rebuilding destroyed and impoverished areas ("Hamedani," 2015). In April 2015, Iran's cultural attaché to Syria met with the heads of several Syrian NGOs in Damascus to discuss ways in which they could learn from Iranian institutions, such as the Sisters Basij, that have three decades of experience helping people cope with the effects of war and assisting the families of martyrs ("Didaar-e Azaa-ye," 2015).

Conclusion

Iran has shown a strong intent and capability to employ political warfare as a means of increasing its power in the Middle East. Tehran's strategy ultimately is motivated by the fear that the West and its allies intend to overthrow the Islamic Republic, and the understanding that Iran must engage in unconventional measures to compensate for its military disadvantage.

As illustrated by its activities in Iraq and Syria, Iran likely will continue to exploit regional instability to establish proxy militias composed of recruits from disenfranchised populations with the aim of using them both to pressure politicians and to become political actors in their own right. Iran will seek to inspire the region's inhabitants with its revolutionary ideology, while attempting to co-opt grassroots movements pursuing democratic reform. It likely also will go after market share in weak economies as a means of exerting political pressure over dependent governments.

Iran's political warfare strategy has been largely successful in Iraq. Tehran has succeeded in limiting U.S. power, creating loyal political parties and militias, and posing a challenge to the Iraqi clergy's influence. The rise of ISIL has allowed Iran to strengthen its hold even further over Iraqi Shi'a militias and the Baghdad government. In the face of weak Iraqi institutions, some predict that the PMUs could transform eventually into a parallel state under Iran's influence.[29] However, the Islamic Republic will always have to contend with Iraqi nationalism among the general population. Therefore, Tehran's ultimate objective likely is to encourage the degree of sectarian and ethnic fragmentation among Iraqis that will be necessary to block the return of a strong, nationalist Baghdad government, while simultaneously striving to prevent the outbreak of large-scale violence that could spill over into Iran.

Iran has conducted a significant level of political warfare in Syria, but has achieved less success than in Iraq. The lack of a large Shi'a population, combined with the largely secular nature of the regime's Syrian backers, means that popular support for the Islamic Republic's religious ideology is limited. While Tehran's charity work in regime-held areas undoubtedly has created goodwill among the needy, without a history of extensive cultural exchange between Iran and Syria, it is unclear how long their hearts and minds will remain inclined toward a Shi'a theocracy. From a military standpoint, Iran has been effective in bolstering Assad's forces, namely the NDF. But Iranian officials likely do not expect the NDF to provide unequivocal support to Tehran if the Assad government were to fall. Therefore, much of Iran's political warfare efforts have focused on recruiting foreign Shi'a militias to fight

---

[29] Indeed, Iranian officials have depicted the PMU network as an Iraqi version of the Basij, Iran's omnipresent paramilitary force responsible for defending the Islamic revolution against both external and internal enemies.

in Syria. In a postwar situation without Assad, Iran would use these militants to ensure access to Hezbollah's supply routes. The militias could also place pressure on any future Damascus government. While Iran has effectively exploited Syria's isolation to its economic benefit, it remains to be seen whether increased Iranian market share will translate into long-term political influence. An isolated postwar Damascus government likely would be subject to Iranian political machinations. However, if Syria were to reengage with the global economy, Iran's political influence would be diluted.

At the broader regional level, the Islamic Republic's political warfare efforts are confronted by the rise in sectarian tensions, which challenges Tehran's desire to be seen as the defender of all oppressed Muslims. Most damaging to Iran in this regard has been its role, alongside the Shi'a group Hezbollah, in crushing the mainly Sunni opposition to the Assad regime.[30] In some cases, such as in Syria, Iran has fueled the sectarianism through its political warfare. In other instances, such as in Yemen and Bahrain, Iran's Sunni rivals have highlighted sectarian differences to justify their own interventions (Scotten, 2015). Regardless of the cause, as long as the Sunni-Shi'a divide remains at the forefront of Middle Eastern geopolitics, the Islamic Republic's ability to spread its revolutionary message and extend its influence will be limited.

References—Chapter 4

"Aakharin Mosahebe-ye Sardaar Hamedani dar Mowred-e Soorieh va Fetne-ye 88 [Final Interview of General Hamedani on Syria and the 88 Sedition]," *Otagh Khabar*, October 10, 2015. As of January 19, 2017:
http://otaghkhabar24.ir/news/18722

Abdollahian, Hossein Amir, "Panj Nokteh Darbaare-ye Omgh-e Estraatezhik-e Iran [Five Points Regarding Iran's Strategic Depth]," *Office for Preservation and Publication of Works of Imam Ayatollah Khamenei*, April 19, 2013. As of January 19, 2017:
http://farsi.khamenei.ir/others-note?id=22404

30 Though in a 2006 Zogby poll of regional attitudes, Iran received a 75 percent favorability rating—largely due to its anti-Israel stance—by 2011, Iran's favorability among Arabs had seen a drastic 60–70 point drop ("The Rise and Fall," 2013; Zogby, 2014).

Abedin, Mahan, "Iran Breaks New Ground in Iraq," *Middle East Eye*, May 29, 2015. As of January 19, 2017:
http://www.middleeasteye.net/columns/iran-breaks-new-ground-iraq-389206135

"The Ailing Ayatollah," *The Economist*, December 5, 2015. As of January 19, 2017:
http://www.economist.com/news/middle-east-and-africa/21679508-ali-al-sistani-iraqs-best-hope-curbing-iranian-influence-he-85-and?fsrc=scn/tw/te/pe/ed/theailingayatollah

Al-Jaffal, Omar, "Iraqi Shiites Join Syria War," *Al-Monitor*. October 29, 2013. As of January 23, 2017:
http://www.al-monitor.com/pulse/originals/2013/10/iraqi-shiites-join-syrian-war.html

Al-Kifaee, Fadel, "Between Najaf and Tehran," *Carnegie Endowment for International Peace*, October 9, 2012. As of January 19, 2017:
http://carnegie-mec.org/sada/49629

Al-Salhy, Suadad, "Iraqi Sunnis Join Shia Militias to Fight IS Militants," *Middle East Eye*, June 15, 2015,
http://www.middleeasteye.net/news/iraqi-sunnis-join-shiite-militias-fight-militants-520291754.

Allam, Hannah, Jonathan S. Landay, and Warren P. Strobel, "Iranian Outmaneuvers U.S. in Iraq," *McClatchy Newspapers*, April 28, 2008,
http://www.mcclatchydc.com/2008/04/28/35146/iranian-outmaneuvers-us-in-iraq.html.

Arango, Tim, "Dozens Killed in Battles across Iraq as Sunnis Escalate Protests Against Government," *New York Times*, April 23, 2013,
http://www.nytimes.com/2013/04/24/world/middleeast/clashes-at-sunni-protest-site-in-iraq.html.

"Babil Council Member to Al-Etejah Press: Apaches Land for Unknown Reason," *Al-Etejah Press*, February 20, 2015. As of January 19, 2017:
http://aletejahtv.org/index.php/permalink/47558.html (Arabic).

Black, Ian, "How Iran's Shadowy Role in Syria Fuels Paranoia and Wariness," *The Guardian*, September 21, 2015. As of January 20, 2017:
http://www.theguardian.com/world/2015/sep/21/irans-shadowy-influence-in-syrias-maelstrom-fuels-paranoia-and-wariness

———, "Iran to Launch Spanish-Language Television Channel," *The Guardian*, September 30, 2010. As of January 20, 2017:
http://www.theguardian.com/world/2010/sep/30/iran-spanish-language-television

Boldaji, Cyrus Barna, "Entekhabaat-e Araagh va Entezaraat-e Iran [Iraq's Elections and Iran's Expectations]," *IR Diplomacy*, March 7, 2010. As of January 20, 2017:
http://www.irdiplomacy.ir/fa/page/11418

Bozorgmehr, Najmeh, "Iran Eyes Iraq Trade Decades after War," *Financial Times*, October 28, 2015. As of January 20, 2017:
http://www.ft.com/intl/cms/s/0/668edf26-78cd-11e5-933d-efcdc3c11c89.html

Brennan, Rick, Jr., Charles P. Ries, Larry Hanauer, Ben Connable, Terrence K. Kelly, Michael J. McNerney, Stephanie Young, Jason H. Campbell, and K. Scott McMahon, *Ending the U.S. War in Iraq: The Final Transition, Operational Maneuver, and Disestablishment of United States Forces-Iraq*, Santa Monica, Calif.: RAND Corporation, 2013.

Chulov, Martin, "Controlled by Iran, the Deadly Militia Recruiting Iraq's Men to Die in Syria," *The Guardian*, March 12, 2014. As of January 20, 2017:
http://www.theguardian.com/world/2014/mar/12/iraq-battle-dead-valley-peace-syria

Comemoração do Eid al Fiter em Foz do Iguaçu [Commemoration of Eid al-Fitr in Foz do Iguaçu], *Sociedade Beneficente Islâmica*, July 28, 2014. As of February 6, 2017:
http://www.islamfoz.com.br/comemoracao-eid-al-fiter-foz-iguacu

"Dafaatar-e Rayzani-ye Farhangi-ye Iran dar Karbala va Najaf Raah Andaazi Mishavad [Iranian Cultural Centers to Be Started in Karbala and Najaf]," *Islamic Republic News Agency*, May 25, 2015,
http://www.irna.ir/fa/News/81621194.

Dagher, Sam, "Devotion and Money Tie Iranians to Iraqi City," *New York Times*, May 30, 2009. As of January 20, 2017:
http://www.nytimes.com/2009/05/31/world/middleeast/31karbala.html.

Daragahi, Borzou, "Iran's Intervention in Syria Is Bold, Unprecedented—And Possibly a Disaster," *BuzzFeed News*, November 16, 2015. As of January 20, 2017:
http://www.buzzfeed.com/borzoudaragahi/inside-irans-bold-unprecedented-intervention-in-syria#.qnP3NL3KQ

———, "Iraq: The Lesser of Two Evils." *Financial Times*, April 6, 2015,
http://www.ft.com/intl/cms/s/0/2d446228-d79b-11e4-94b1-00144feab7de.html#axzz3tOz35SVy.

Daragahi, Borzou, Erika Solomon, Najmeh Bozorgmehr, and Geoff Dyer, "Battle for Iraq: The Iranian Connection," *Financial Times*, November 10, 2015. As of January 20, 2017:
http://www.ft.com/cms/s/0/18136ee6-68c2-11e4-af00-00144feabdc0.html.

Dehghan, Saeed Kamali, "Afghan Refugees in Iran Being Sent to Fight and Die for Assad in Syria," *The Guardian*, November 5, 2015,
http://www.theguardian.com/world/2015/nov/05/iran-recruits-afghan-refugees-fight-save-syrias-bashar-al-assad.

Dehghanpisheh, Babak, "Iran Recruiting Pakistani Shiite Fighters for Combat in Syria," Reuters, December 11, 2015. As of January 20, 2017: http://www.dailystar.com.lb/News/Middle-East/2015/Dec-11/326894-iran-recruiting-pakistani-shiite-fighters-for-combat-in-syria.ashx

"Didaar-e Azaa-ye Karavaan-e Zaaeraan-e Solh ba Mofti-ye Azame- Soorieh [Peace Pilgrim Caravan's Visit with Syria's Grand Mufti]," *International Union of Unified Ummah*, April 14, 2014. As of Janauary 20, 2017: http://iuuu.ir/fa/7757

"Didaar-e Vaabaste-ye Farhangi-ye Iran ba Roasa-ye Moasesaat-e Mardom Nahaad-e Soorieh [Iran Cultural Attache's Visit with Heads of Syrian NGOs]," *Islamic Culture and Relations Organization*, April 19, 2015. As of January 20, 2017: http://www.icro.ir/index.aspx?siteid=261&pageid=11751&newsview=628941

Eisenstadt, Michael, *Martyrdom, Victory, and Expediency in the Decisionmaking of the Islamic Republic of Iran: Implications for Deterrence and Warfighting*, PowerPoint Presentation.

———, *The Missing Lever: Information Activities against Iran*, Washington, D.C.: The Washington Institute for Near East Policy, 2010.

Eisenstadt, Michael, Michael Knights, and Ahmed Ali, *Iran's Influence in Iraq: Countering Tehran's Whole-of-Government Approach*, Washington, D.C.: The Washington Institute, 2011.

"Enemy's Infiltration Big Threat to Iran: Supreme Leader," *Ahlul Bayt News Agency*, September 16, 2015. As of January 20, 2017: http://en.abna24.com/service/iran/archive/2015/09/16/711025/story.html

"Ersaal-e Komakha-ye Gheir-e Naghdi beh Mardom-e Soorieh be Arzesh-e 3 Miliaard Riaal [3 Billion Rials in Non-Cash Aid Sent to People of Syria]," *ANAJ News*, October 28, 2015. As of January 20, 2017: http://www.anaj.ir/News/25567

"Farmaande-ye Sepah: 100 Hezar Niroo-ye Mardomi Dar Soorieh Sazmaandahy Shodand [IRGC Commander: 100,000 People's Forces Organized in Syria]," *IR Diplomacy*, October 18, 2015. As of January 20, 2017: http://www.irdiplomacy.ir/fa/page/1952987

Felter, Joseph, and Brian Fishman, *Iranian Strategy in Iraq: Politics and "Other Means,"* West Point, N.Y.: Combatting Terrorism Center, 2008.

"Firoozabadi: Hamkaari-ye Iran va Amrika Hargez Etefaagh Nemioftad [U.S.-Iran Cooperation will Never Happen]," *Fars News Agency*, June 18, 2014. As of January 20, 2017: http://www.farsnews.com/newstext.php?nn=13930328001312

Fulton, Will, Joseph Holliday, and Sam Wyer, *Iranian Strategy in Syria*, Washington, D.C.: Institute for the Study of War, 2013.

Garcia, Eduardo, and Saul Hudson, "Iranian President Courts Latin American Leftists," Reuters, September 27, 2007. As of January 20, 2017:
http://www.reuters.com/article/us-bolivia-iran-idUSN2734139720070928

Ghazi, Yasir, and Tim Arango, "Iraqi Sects Join Battle in Syria on Both Sides," *New York Times*, October 27, 2012. As of January 20, 2017:
http://www.nytimes.com/2012/10/28/world/middleeast/influx-of-iraqi-shiites-to-syria-widens-wars-scope.html?_r=0

Gholamzadeh, Hamid Reza, "Soft War vs. Soft Power," *Mehr News*, October 14, 2015. As of January 20, 2017:
http://en.mehrnews.com/news/111041/Soft-War-vs-Soft-Power

Gholipour, Behnam, "Deaths in Iraq Show Two Sides of Iran's Role in Sectarian Conflict," *The Guardian*, December 30, 2014. As of Janaury 20, 2017:
http://www.theguardian.com/world/iran-blog/2014/dec/30/iran-militia-leaders-killed-iraq-battat-taqavi

Golshan-Pezhou, Mahmoud Reza, "Lozoom beh Rokh Keshi-ye Ghodrat-e Narm-e Iran [Need for Boosting Iran's Soft Power]," *IR Diplomacy*, March 30, 2012. As of January 20, 2017:
http://www.irdiplomacy.ir/fa/page/1899414

Habib, Mustafa, "A House Divided: Iranian, Iraqi Ayatollahs Square Off in Baghdad Streets," *Informed Comment*, September 4, 2015. As of January 20, 2017:
http://www.juancole.com/2015/09/ayatollahs-baghdad-streets.html

———, "Iraq's Controversial Shiite Militias Fight Among Themselves," *Niqash*, June 18, 2015. As of January 20, 2017:
http://www.niqash.org/en/articles/politics/5033

———, "Senior Iraqi Cleric 'Saves' the Government—and Iraq from Iran?" *Niqash*, August 12, 2015. As of January 20, 2017:
http://www.niqash.org/en/articles/politics/5077

———, "Shifting Shiite Allegiances Change Political Landscape," *Niqash*, March 15, 2012. As of January 20, 2017:
http://www.niqash.org/en/articles/politics/3011

"Hadaf az Saakht-e Film-e Avaazha-ye Sarzamin-e Man Rowshangari Baraye Mardom-e Jahaan Bood [Goal for Creating Film Songs of My Homeland Was Enlightening People of the World]," *Aviny Film*, June 29, 2015. As of January 20, 2017:
http://avinyfilm.ir/content/12943

Haghighatkha, Mehdi, "Omgh-e Raahbordi-ye Iran: Baghdad ta Karakas [Iran's Strategic Depth: From Baghdad to Caracas]," *Borhan News*, August 5, 2013. As of January 20, 2017:
http://borhan.ir/NSite/FullStory/News/?Id=5630

"Hamedani: Basij-e Sazandegi dar Soorieh Tashkil Shod [Hamedani: Construction Basij Established in Syria]," *Mashregh News*, March 1, 2015. As of January 20, 2017:
http://www.mashreghnews.ir/fa/news/393797

Harsin, Hossein, Mojtaba Tavisarkani, and Leila Jafari, "Geopolitik-e Ghodrat-e Narm-e Iran [Geopolitics of Iran's Soft Power]," *Oloom-e Siaasi*, Vol. 4, No. 2 (Spring 2009), 225–69.

Hashem, Ali, "Iraq's Shiite Forces Claim Victory over IS," *Al-Monitor*, March 16, 2015. As of January 20, 2017:
http://www.al-monitor.com/pulse/originals/2015/03/iraq-shiite-hezbollah-nujaba-victory-islamic-state.html

Healy, Jack, "Arrest Order for Sunni Leader in Iraq Opens New Rift," *New York Times*, December 19, 2011. As of January 20, 2017:
http://www.nytimes.com/2011/12/20/world/middleeast/iraqi-government-accuses-top-official-in-assassinations.html?_r=0

Hiltermann, Joost, "Iraq: The Clerics and the Militias," *New York Review of Books*, October 13, 2015. As of January 20, 2017:
http://www.nybooks.com/daily/2015/10/13/iraq-clerics-militias

Hossein, Saeed, and Sohrab Anami Alamdari, "Zarfiyatha-ye Ghodrat-e Narm-e Jomhouri-ye Eslami-ye Iran dar Moghaabeleh ba Tahdidaat-e Narm-e Amrika [Islamic Republic of Iran's Soft Power Capacities to Counter America's Soft Threats]," *Gozaaresh-e Jomhour*, No. 26, March 2008, pp. 3–28.

Ilyas, Shayerah, *Iran's Economic Conditions*, Washington, D.C.: Congressional Research Services, 2009.

"International Quds Day Celebration," *Sierra Express Media*, August 7, 2013. As of January 23, 2017:
http://www.sierraexpressmedia.com/?p=59726

"Iran Quds Chief Visited Russia Despite U.N. Travel Ban," Reuters, August 7, 2015. As of January 20, 2017:
http://www.reuters.com/article/us-russia-iran-soleimani-idUSKCN0QC1KM20150807#THalTZdtqtlu3QeK.97

"Iran Sending Thousands of Afghans to Fight in Syria," *Human Rights Watch*, January 29, 2016. As of January 20, 2017: https://www.hrw.org/news/2016/01/29/iran-sending-thousands-afghans-fight-syria

"Iran Set Up Terrorist Networks in Latin America," *The Jerusalem Post*, May 30, 2013. As of January 20, 2017:
http://www.jpost.com/Iranian-Threat/News/Prosecutor-Iran-set-up-terrorist-networks-in-Latin-America-314793

"Iraq's Child Soldiers," *Al-Monitor*, August 20, 2015. As of January 20, 2017:
http://www.al-monitor.com/pulse/originals/2015/08/iraq-iran-child-soldiers.html

"Iraq's Most Fearsome Shia Militia Leader Speaks Out," *The Daily Beast*, August 24, 2015. As of January 20, 2017:
http://www.thedailybeast.com/articles/2015/08/24/iraq-s-most-fearsome-shia-militia-leader-speaks-out.html

Kajjo, Sirwan, "Hezbollah Expands Military Presence in Syrian War," *Voice of America*, November 10, 2015. As of January 20, 2017:
http://www.voanews.com/content/hezbollah-expands-military-presence-syrian-war/3052160.html

Kenner, David, "For God and Country, and Iran," *Foreign Policy*, March 5, 2015. As of January 20, 2017:
http://foreignpolicy.com/2015/03/05/for-god-and-country-and-iran

Khalaji, Mehdi, *The Last Marja: Sistani and the End of Traditional Religious Authority in Shiism*, Washington, D.C.: Washington Institute for Near East Policy, 2006.

Khalkhali, Ali Mousavi, "Cheraa Safar-e Al-Abadi beh Tehran Mohem Ast? [Why Is Al-Abadi's Trip to Tehran Important?]," *IR Diplomacy*, October 20, 2014, http://www.irdiplomacy.ir/fa/page/1939790

Khamenei, Ali, "Bayanaat Dar Didar-e Farmandehaan-e Sepah-e Pasdaraan-e Enghelaab-e Eslami [Statements During Visit with Commanders of Islamic Revolutionary Guards Corps]," *Office for Preservation and Publication of Works of Imam Ayatollah Khamenei*, September 16, 2015,
http://farsi.khamenei.ir/keyword-content?id=1034

———, "Tahlil-e Vaziat-e Mantaghe [Analysis of the Regional Situation]," *Office for Preservation and Publication of Works of Imam Ayatollah Khamenei*, November 26, 2015. As of January 20, 2017:
http://farsi.khamenei.ir/photo-album?id=31524

Khamenei.ir, "Massive election turnout guarantees the security of the country," January 9, 2016. As of January 20, 2017:
http://english.khamenei.ir/news/3122/Massive-election-turnout-guarantees-the-security-of-the-country

Knights, Michael, "Iran's Influence in Iraq: Game, Set But Not Match to Iran," *The Guardian*, October 17, 2010. As of January 20, 2017:
http://www.theguardian.com/world/2010/oct/17/iran-influence-iraq-tehran

"Komakha-ye Kheirie-ye Daneshjooyaan-e Irani be Niazmandaan va Avaregaan-e Soori Tahavol Shod [Iranian University Students' Aid and Charity Transferred over to Needy and Displaced Syrians]," *Tasnim News*, October 27, 2014. As of January 20, 2017:
http://www.tasnimnews.com/fa/news/1393/08/05/541339

"Komakha-ye Kheirieh-ye Iran beh Soorieh Hargez Motevaghef Nashod [Iran's Charity Aid to Iran Never Stopped]," *Fars News Agency*, September 6, 2015. As of January 20, 2017:
http://www.farsnews.com/newstext.php?nn=13940615000233

Lamothe, Dan, "Qassem Soelimani, Iran's Spy Commander, Introduced to a New Generation," *Washington Post*, March 5, 2015. As of January 20, 2017: https://www.washingtonpost.com/news/checkpoint/wp/2015/03/05/qassem-suleimani-irans-spy-commander-introduced-to-a-new-generation

"Lebanon's Hizbollah Turns Eastward to Syria," *International Crisis Group Middle East Report*, No. 153, May 27, 2014.

Levitt, Matthew, and Phillip Smyth, "Kataib al-Imam Ali: Portrait of an Iraqi Shiite Militant Group Fighting ISIS," *The Washington Institute for Near East Policy*, January 5, 2015. As of January 20, 2017:
http://www.washingtoninstitute.org/policy-analysis/view/kataib-al-imam-ali-portrait-of-an-iraqi-shiite-militant-group-fighting-isis

Lucas, Scott, "Iran's Economic Stakes in the Assad Regime," *EA Worldview*, June 7, 2015. As of January 20, 2017:
http://eaworldview.com/2015/06/syria-analysis-irans-economic-stakes-in-the-assad-regime

Malas, Nour, "Iraqi Premier's Rule Again Questioned," *Wall Street Journal*, September 23, 2015. As of January 20, 2017:
http://www.wsj.com/articles/iraqi-premiers-rule-again-questioned-1443050315

Maltzahn, Nadia von, "The Case of Iranian Cultural Diplomacy in Syria," *Middle East Journal of Culture and Communication*, Vol. 2, 2009, pp. 33–50.

———, "Iran's Cultural Diplomacy," Chapter Twelve in *Iran and the Challenges of the Twenty-First Century: Essays in Honour of Mohammad-Reza Djalili*, ed. H. E. Chehabi, Farhad Khosrokhavar, and Clement Therme, Costa Mesa: Mazda Publishers, 2013, pp. 205–21.

———, *The Syria Iran Axis: Cultural Diplomacy and International Relations in the Middle East*, London: I. B. Tauris, 2013.

Mamouri, Ali, "Shiite Seminaries Divided on Fatwas for Syrian Jihad," *Al-Monitor,* July 29, 2013. As of January 20, 2017:
http://www.al-monitor.com/pulse/en/originals/2013/07/syria—jihad-fatwas-shiite-clergy-iran-iraq.html

Morris, Loveday, "Shiite Militias in Iraq Begin to Remobilize," *Washington Post*, February 9, 2014. As of January 20, 2017:
https://www.washingtonpost.com/world/middle_east/shiite-militias-in-iraq-begin-to-remobilize/2014/02/09/183816c6-8f59-11e3-878e-d76656564a01_story.html

Nader, Alireza, *Iran's Role in Iraq: Room for Cooperation?* Santa Monica, Calif.: RAND Corporation, 2015.

Namaa, Kamal, "Fighting Erupts as Iraq Police Break Up Sunni Protest Camp," Reuters, December 30, 2013. As of January 20, 2017:
http://www.reuters.com/article/us-iraq-violence-idUSBRE9BT0C620131230#GAg6T0AVqsEMHIyj.97

Nasr, Vali, *The Shia Revival: How Conflicts within Islam Will Shape the Future*, New York: W. W. Norton & Company, 2007.

Nasseri, Ladane, and Golnar Motevalli. "Investing in Iran? You'd Better Like Tea, Cake and Bureaucracy," *Bloomberg Business*, October 5, 2015. As of January 20, 2017:
http://www.bloomberg.com/news/articles/2015-10-05/investing-in-iran-you-d-better-like-tea-cake-and-bureaucracy

Pahlavi, Pierre, "Understanding Iran's Media Diplomacy," *Israel Journal of Foreign Affairs* 6, No. 2, 2012, pp. 21–33.

Parker, Ned, "Interior Ministry Mirrors Chaos of a Fractured Iraq," *New York Times*, July 30, 2007. As of January 20, 2017:
http://articles.latimes.com/2007/jul/30/world/fg-interior30

———, "Power Failure in Iraq as Militants Outgun State," Reuters, October 21, 2015. As of January 20, 2017:
http://www.reuters.com/investigates/special-report/iraq-abadi

"Presidente Italiano Recibe a Su Par Iraní en Roma [Italian President Welcomes His Iranian Counterpart in Rome]," *HispanTV*, January 26, 2016. As of January 20, 2017:
http://hispantv.com/newsdetail/Reporteros/202596/mattarella-recibe-rohani-visita-italia

Pyruz, Mark, "Syrian NDF Modeled on Basij," *Uskowi on Iran*, March 28, 2016. As of January 20, 2017:
http://hispantv.com/newsdetail/Reporteros/202596/mattarella-recibe-rohani-visita-italia

Qaidaari, Abbas, "Comparing Iraq's Shiite Forces to Iran's Basij," *Al-Monitor*, May 11, 2015. As of January 20, 2017:
http://www.al-monitor.com/pulse/originals/2015/05/iran-iraq-iraq-basij-forces.html

Remnick, David, "Negotiating the Whirlwind: Can Secretary of State John Kerry Break through in Syria?" *New Yorker*, December 21, 2015. As of January 20, 2017:
http://www.newyorker.com/magazine/2015/12/21/negotiating-the-whirlwind

Rezakhah, Alireza, "Mokhtasaat-e Nofooz-e Mantaghei-ye Iran Dar Moj-e Sevom-e Tahavolaat-e Beinolmelal [Particulars of Iran's Regional Influence amid Third Wave of International Transformations]," *Office for Preservation and Publication of Works of Imam Ayatollah Khamenei*, January 17, 2015. As of January 20, 2017:
http://farsi.khamenei.ir/others-note?id=28694

"The Rise and Fall of Iran in Arab and Muslim Eyes: A New Poll," *Project on Middle East Democracy*, March 6, 2013. As of January 20, 2017:
http://www.aaiusa.org/pomed-notes-the-rise-and-fall-of-iran-in-arab

Roozbahani, Mehdi, "Diplomaasi beh Sabk-e Irani dar Karbala [Iranian-Style Diplomacy in Karbala]," *Tabnak*, December 15, 2015. As of January 20, 2017: https://www.tabnak.ir/fa/news/458284

Rosenfeld, Jesse, "Hezbollah Fighters Are Fed Up with Fighting Syria's War," *The Daily Beast*, December 30, 2015. As of January 20, 2017: http://www.thedailybeast.com/articles/2015/12/30/hezbollah-fighters-are-fed-up-with-fighting-syria-s-war.html

"Sanaa Is the Fourth Arab Capital to Join the Iranian Revolution," *Middle East Monitor*, September 27, 2014. As of January 20, 2017: https://www.middleeastmonitor.com/news/middle-east/14389-sanaa-is-the-fourth-arab-capital-to-join-the-iranian-revolution

Scotten, Ali G., "Countering the Sunni-Shia Divide," *Foreign Policy Association*, May 14, 2015. As of January 20, 2017: http://foreignpolicyblogs.com/2015/05/14/countering-the-sunni-shia-divide

———, "Iran Picks New Battle Against Islamic State with State-Commissioned Film," *The Guardian*, November 24, 2015. As of January 20, 2017: http://www.theguardian.com/world/iran-blog/2015/nov/24/iran-isis-movie-arabs-songs-of-my-homeland

Shabani, Mohammad Ali, "The Impending Shiite Leadership Crisis Baghdad Doesn't See Coming," *Al-Monitor*, August 26, 2015. As of January 20, 2017: http://www.al-monitor.com/pulse/originals/2015/08/iran-iraq-politics.html

"Shadowy Role in Syria Fuels Paranoia and Wariness," *Guardian*, September 21, 2015. As of January 20, 2017: http://www.theguardian.com/world/2015/sep/21/irans-shadowy-influence-in-syrias-maelstrom-fuels-paranoia-and-wariness

"Sistani Questions Qassem Soleimani's Influence," *NOW*, September 8, 2015. As of January 20, 2017: https://now.mmedia.me/lb/en/NewsReports/565861-sistani-questions-qassem-soleimanis-influence-report

Slackman, Michael, "Iraqi Ties to Iran Create New Risks for Washington," *New York Times*, June 8, 2006. As of January 20, 2017: http://www.nytimes.com/2006/06/08/world/middleeast/08iran.html?pagewanted=print&_r=0

Sly, Liz, "Iraqis Think the U.S. Is in Cahoots with the Islamic State, and It Is Hurting the War," *Washington Post*, December 1, 2015. As of January 20, 2017: https://www.washingtonpost.com/world/middle_east/iraqis-think-the-us-is-in-cahoots-with-isis-and-it-is-hurting-the-war/2015/12/01/d00968ec-9243-11e5-befa-99ceebcbb272_story.html

Smyth, Phillip, "Iran-Backed Iraqi Militias Are Pouring into Syria," *Business Insider*, October 2, 2015. As of January 20, 2017: http://www.businessinsider.com/iran-backed-iraqi-militias-are-pouring-into-syria-2015-10

———, *The Shiite Jihad in Syria and Its Regional Effects*, Washington, D.C.: The Washington Institute for Near East Policy, 2015.

Solomon, Erika, "Syrian Government Guerilla Fighters Being Sent to Iran for Training," April 4, 2013. As of January 20, 2017:
http://www.reuters.com/article/us-syria-iran-training-insight-idUSBRE9330DW20130404#GrKw9L526Dc18UAs.97

Solomon, Jay, and Dion Nissenbaum, "With Maliki out of the Picture, U.S. Prepares to Boost Iraq Aid," *Wall Street Journal*, August 15, 2014. As of January 20, 2017:
http://www.wsj.com/articles/with-maliki-out-of-the-picture-u-s-prepares-to-boost-iraq-aid-1408143896

Stanford University, "Mahdi Army," *Mapping Militant Organizations*. As of November 10, 2014:
http://web.stanford.edu/group/mappingmilitants/cgi-bin/groups/view/57

Steele, Jonathan, "Sunni Tribes Joining Shia Militias as War Against IS Heats Up in Iraq," *Middle East Eye*, December 1, 2015. As of January 20, 2017:
http://www.middleeasteye.net/news/sunni-tribes-joining-shia-militias-war-against-heats-iraq-1175770052

"Street Battles Between Former and Current Terrorist Militias," *Niqash*, September 26, 2013. As of January 20, 2017:
http://www.niqash.org/en/articles/politics/3301

Sullivan, Marisa Cochrane, "Iran's Soft Power in Iraq," *Iran Tracker*, August 4, 2009. As of January 20, 2017:
http://www.irantracker.org/analysis/iran%E2%80%99s-soft-power-iraq

Szrom, Charlie, *Ahmadinejad in West Africa: What Iranian Outreach to the Region Reveals about Tehran's Foreign Policy*, Washington, D.C.: American Enterprise Institute, 2010.

"Taazeh-tarin Sokhanaan-e Sardaar Hamedani Dar Mored-e Soorieh [General Hamedani's Latest Statemnts on Syria]," *Islamic Students News Agency*, June 27, 2014. As of January 17, 2017:
http://isna.ir/fa/news/93040603516

"Tahghighaat-e Diin va Siyaasat dar Amrikaa-ye Laatin [Studies of Religion and Politics in Latin America]," *Pegah Domains*, May 1, 2010. As of January 23, 2017:
http://www.abp-shiaworld.com/fa.php/page,3408A48968.html

"Tahlil-e Rahbar-e Enghelaab az Havades-e Akhir-e Mantaghe [Supreme Leader's Analysis of Recent Regional Incidents]," *Office for Preservation and Publication of Works of Imam Ayatollah Khamenei*, July 1, 2014. As of January 20, 2017:
http://farsi.khamenei.ir/speech-content?id=26839

"Tashkiil-e Goroohi Vizhe dar Soorieh Baraaye Moghabeleh ba Eghdaamaat-e Tafreghe Afkanaane-ye Doshman [Establishment of Special Group in Syria to Combat Enemy's Divisive Actions]," *Defa Press*, March 6, 2015. As of January 20, 2017:
http://www.defapress.ir/Fa/News/42615

"Tashkil-e Setaad-e Poshtibaani az Jebhe-ye Moghaavemat-e Soorieh dar Malayer [Creation of Resistance Front Support Staff in Malayer]," *Hamedan Basij News Organization*, May 15, 2014. As of January 20, 2017:
http://www.freezepage.com/1400155869ESQWLNMULB

Tavasoli, Hossein, "Dastaavardha-ye Nofooz-e Iran Dar Mantagheh [Results of Iran's Influence in the Region]," *Fars News Agency*, January 25, 2015. As of January 20, 2017:
http://www.farsnews.com/printable.php?nn=13931104001409

Thaler, David E., Alireza Nader, Shahram Chubin, Jerrold D. Green, Charlotte Lynch, and Frederic Wehrey, *Mullahs, Guards, and Bonyads: An Exploration of Iranian Leadership Dynamics*, Santa Monica, Calif.: RAND, 2010.

Vaezi, Mahmoud, "Roykard-e Gorooha-ye Daakheli-ye Araagh beh Iran va Amrika [Domestic Iraqi Groups' Approaches to Iran and the U.S.]," in *Post-Saddam Iraq and Regional Players*. As of January 20, 2017:
http://www.csr.ir/departments.aspx?abtid=07&depid=44&semid=1464iran. Tehran: Expediency Discernment Council Center for Strategic Research, 2008

Visser, Reidar, "Religious Allegiances Among Pro-Iranian Special Groups in Iraq," *CTC Sentinel*, Vol. 4, No. 9, September 2011, pp. 5–8.

Wastnidge, Edward, "The Modalities of Iranian Soft Power: From Cultural Diplomacy to Soft War," *Politics*, December 2014, pp. 1–14.

"Why Iraq's Shiites Are Taking Jihad to Syria and Supporting Al-Assad," *Niqash*, June 25, 2013,
http://www.niqash.org/en/articles/politics/3260.

Wyer, Sam, "The Resurgence of Asa'ib Ahl Al-Haq," *Middle East Security Report*, December 7, 2012.

Yazigi, Jihad, "Iran Building Long-Term Influence in Syria," *Middle East Eye*, May 12, 2015. As of January 20, 2017:
http://www.middleeasteye.net/columns/iran-building-long-term-influence-syria-1770733224

Zambelis, Chris, Personal communication, November 8, 2015.

Zogby, James, "The Arab and Iranian Disconnect," *Huffington Post*, November 29, 2014. As of January 20, 2017:
http://www.huffingtonpost.com/james-zogby/the-arab-and-iranian-disconnect_b_6240264.html

CHAPTER FIVE

Case Study: ISIL

Introduction

The Islamic State of Iraq and the Levant has broken new ground for a terrorist group in achieving substantial reach through both military and nonmilitary means. The self-proclaimed Islamic caliphate not only holds vast swaths of territory in Syria and Iraq, but has employed sophisticated methods of governance and extortion, incorporated affiliates as provinces in countries as far-flung as Tunisia and Afghanistan, and pioneered an innovative, dispersed model of propaganda production and dissemination. This chapter will focus on those practices related to political warfare and in particular its innovative use of social media and other communications methods.

We begin this case study by examining ISIL's theoretical and practical approaches to political warfare by means of analyzing ISIL strategy and internal documents, followed by an examination of how ISIL has applied these principles, according to a rearranged DIME construct. First, ISIL's diplomatic practices include its use of affiliates and alliances, such as proxies and allied militant groups, as well as its occasional cooperation with some regional governments. Second, ISIL attempts to establish economic dominance through controlling independent sources of critical resources within its territory and exploiting smuggling routes for the black market oil trade with neighboring countries. Third, ISIL utilizes extensive intelligence and counterintelligence systems as well as irregular warfare and atrocities for maximum psychological effect. And finally, ISIL's multilayered information operations campaign incorporates the above efforts, in addition to a targeted online recruitment campaign, into its vast network of propaganda production and dissemination.

We then perform a deep dive to illuminate how ISIL's English-language propaganda differs from its Arabic-language propaganda, in order to better understand how ISIL targets different messages to different audiences. The intent is to contribute to a better U.S. understanding of how a nonstate actor with global ambitions makes appeals across national and linguistic lines. This particular method of lexical analysis, enabled by RAND proprietary software, quantitatively compares English and Arabic propaganda text assembled from ISIL media releases in English and Arabic. The findings indicate that ISIL employs violence and emotive language liberally in Arabic media productions, likely with the aim of mobilizing rank-and-file members, while displaying a more international and marginally more restrained approach in its English-language propaganda. This wider lens appears to serve the dual purpose of intimidating and provoking its enemies while also appealing to more moderate potential supporters who are more interested in the utopian aims of ISIL than in violence. These disparities underscore the inadvisability of focusing exclusively on either English or Arabic editions of ISIL's propaganda; only a multilingual approach can capture a holistic view of ISIL's message.

Finally, the study presents insights gleaned from analyzing the strengths and vulnerabilities of ISIL's multilayered information operations campaign. ISIL's strengths include its uncompromising posture as well as its embrace of structured decentralization, the latter of which might be adopted by ISIL's opponents. However, while ISIL's rebranding of itself as a caliphate has proved a tremendous boon for its recruitment, it also represents a message of vulnerability; selling itself as a caliphate entangles ISIL in promises that it cannot keep.

ISIL's Strategy

ISIL's information operations campaign consists of three interlocking main lines of effort: recruitment, which is the overarching objective of its information operations; governance, which includes a sophisticated intelligence apparatus; and media, which serves to amplify the reach of these efforts as well as multiply the psychological effects of ISIL's brutally irregular warfare. ISIL's economic strategy concentrates on establishing independent sources of money, weapons, oil, gas, and food as a key element of ISIL's attempt to gain leverage against its opponents. While the extent to which ISIL views enduring agreements with foreign nations permissible remains ambiguous, foundational ISIL documents do acknowledge the potential value of temporary treaties.

This review of ISIL's overall strategy relies on a number of primary source documents, including issues of *Dabiq*, ISIL's long-form propaganda magazine in which it often presents detailed justifications of controversial actions; *The Management of Savagery*, written over ten years ago by al Qaeda sympathizer Abu Bakr Naji and reportedly drawn upon by ISIL; and the private notes of ISIL architect Haji Bakr, seized after his death. However, the review draws most heavily from a leaked internal strategy document written by Abu Abdullah Al-Masri in the latter half of 2014 entitled "Principles in the Administration of the Islamic State." This document, reportedly intended as a "blueprint" to train "cadres of administrators" and written in the months following ISIL's declaration of a caliphate, offers valuable insight into high-level ISIL guidance on political warfare concepts (Malik, 2015).

The "Principles" document suggests that ISIL realized it needed to overhaul its political warfare efforts in order to compete with the United States. The document argues that the United States had previously been able to destroy the Islamic State of Iraq (ISI) by founding the Sunni Awakening tribal forces and "portraying it as a treacherous terrorist state of hypocritical political projects." In other words, the United States had won by using affiliates and luring away ISI support in a propaganda war. The document reports that ISIL has "learned from the prior leaders of the state," and intends to "benefit from prior mistakes," approaching political warfare in a more intentional way.

Information Operations

The "Principles in the Administration of the Islamic State" strategy document outlines the three main prongs of the information operations campaign: recruitment, governance, and media. ISIL's recruitment campaign is founded upon the declaration of a caliphate and the formation of training camps. ISIL's governance plan includes a sophisticated information and intelligence apparatus designed to help ISIL expand and maintain control over territory. Finally, ISIL's well-crafted media system empowers its recruitment and governance efforts, while strengthening its embrace of atrocities that are intended to deter opponents and energize supporters with maximum psychological impact.

ISIL's first and most fundamental line of effort within the arena of information operations is recruitment. Key elements include the establishment of indoctrination and training camps and the declaration of a caliphate. In ISIL's "Principles" document, the declaration of a caliphate is described primarily as a rebranding opportunity and recruitment boon.

The declaration of the caliphate could not have been solely motivated by a pragmatic desire to boost recruitment, however; at least some leaders likely felt a sincere religious obligation to reestablish the caliphate. This religious duty is naturally the objective cited in ISIL's propaganda. When Al-Baghdadi gave his first sermon as ISIL's self-declared caliph, he called the caliphate's establishment "the duty of Muslims, which has been lost for centuries . . . and neglected by many Muslims" ("ISIS Abu Bakr al-Baghdadi First First Friday Sermon as So-Called 'Caliph,'" 2014).

However, the internal strategy "Principles" document declares that ISIL's foundational aim in establishing the caliphate was finding "the means to preserve the revolutionaries of Syria on the jihadi trend." It goes on to say that "the announcement of the caliphate was the result of the *mujahideen*'s realizing the lack of advantage in fighting against the idolaters without the existence of a leader and Caliph who could gather the Muslims under his banner." While volunteers had been flooding into Syria to fight the Assad regime, ISIL bemoaned watching these volunteers being diverted into other groups with "private agendas." In order to successfully compete with these other Syrian rebel factions, ISIL assessed that it had to step up its recruiting efforts: "the announcement of the caliphate was the obligation that gathers those arriving in the land of jihad . . . under one banner, one word and one Caliph."

The "Principles" document outlines the establishment of training camps as the second key element of the recruitment process: indoctrination, intended to strip away and replace recruits' previous identities and disincline them from straying. According to the "Principles" document, one such set of camps is designed to induce newly arrived foreign fighters to set "aside local tribalism [and] inclinations of nationalism and ethnic division," making "his affiliation to the religion alone." This involves encouraging the use of Arabic: the "Arabic character should be cultivated . . . in the *muhajideen* [immigrants]" as part of "laying aside the foreign identity that bears in its hidden nature hostility to Islam." The document clarifies that indoctrination is required for both foreign and local recruits, with a "first preparation camp" including lessons on *sharia* (Islamic law) and even a "preparation camp for children," complete with sharia lessons and even "training on bearing light arms." This is followed by an insistence that indoctrination must extend beyond the training camps: "the *mujahid* remains in need of direction and tracking after his completion of the special training session." The document suggests that military commanders, just before they send men into battle, mention "hadiths on the virtue of jihad and endurance . . . following the decisions and instructions of the field commander during the battle [and] the virtue of martyrdom in the path of God."

The drive to increase recruitment has also inspired broader strategizing. An undated ISIL strategy document found in Pakistan, "A Brief History of the Islamic State Caliphate (ISC), The Caliphate According to the Prophet," planned attacks in India and predicted that provoking a full-on U.S. attack would offer the organization another recruitment boon: "Even if the U.S. tries to attack with all its allies, which undoubtedly it will, the *ummah* [Islamic community] will be united, resulting in the final battle" (Carter, 2015).

The second major prong of ISIL's information operations is governance. However, recruitment still remains the core objective underlying ISIL's governance campaign. Governing territory serves two main purposes for ISIL. First, it allows ISIL to reap the propaganda and foreign recruitment benefits of being a caliphate. Second, it allows ISIL to maintain a domestic populace as a recruitment pool independent from foreign attempts to cut off immigration.

ISIL has endeavored mightily to establish the wide-ranging infrastructure of a state. A vast archive of ISIL administrative documents, assembled and translated by Aymenn Jawad Al-Tamimi, outlines a sweeping range of areas of state influence. These include educational plans, textbooks on Islamic doctrine and law for primary school children, the establishment of a "caliphate kindergarten center" as well as the "Central Cub Scouts of the Caliphate Institute," pharmacy regulations, vaccination cards for children, applications for medical school, and exam timetables for nursing colleges. Regulations are required for fishing, receipts for oil and gas, and tax forms for electricity. In May 2015, the Services Department in Mosul went so far as to announce that it was "pleased to invite you to enter the Waritheen Hotel and the Amusement Park for free in relation to the conquest of Wilayat al-Anbar and the conquests in Bilad al-Sham. And praise be to God who supports [and] gives victory to the believers and lowers polytheism and the polytheists" (Archive of Islamic State Administrative Documents, Specimen 4H).

Integrated into ISIL's governance structures is its sophisticated intelligence apparatus, which has allowed ISIL not only to take over but also keep control of occupied territory. When Samir Abd Muhammad al-Khlifawi, the ISIL architect and former intelligence officer under Saddam Hussein known as Haji Bakr, was killed in 2014, a trove of his notes and documents were uncovered, revealing plans for an "Islamic Intelligence State"; these included the sketched-out organizational chart translated in Figure 5.1.

The first purpose for these far-reaching efforts at governance is allowing ISIL to realize the recruitment advantages of being a caliphate. The "Principles" introduction clarifies that the establishment of the Islamic State as a caliphate requires "a comprehensive system" with fully realized concepts. It warns that waging *jihad* and conducting *da'wa*—defending and evangelizing the true religion—are not sufficient functions for a caliphate. Instead, a caliphate requires establishing "an Islamic system of life, a Qur'anic constitution and a system to implement it."

Furthermore, the "Principles" document describes controlling territory as essential for ISIL's attempt to reap the recruitment benefits of statehood. The document even declares, "The state cannot remain

Figure 5.1
Plan for ISIL Intelligence Service Structure

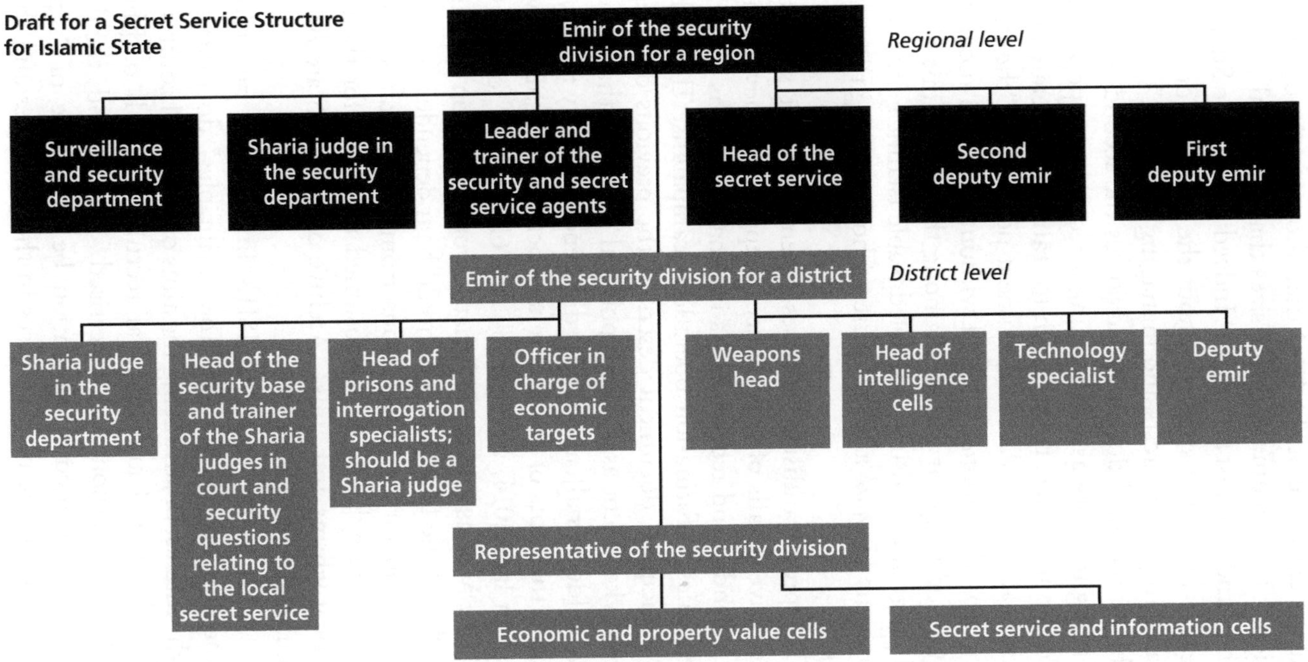

SOURCE: Der Spiegel.
RAND *RR1772-5.1*

without the existence of the land that allows for its continuation and expansion." The document emphasizes that territorial control requires governing land, not just seizing it: "Indeed the Islamic State's seizure of vast areas . . . does not suffice without the existence of an administration managing the interests and managing the crises."

The "Principles" document advances the second primary benefit of governing territory as maintaining a recruitment pool independent from foreign influence, from which to train and generate both soldiers and technical experts. As the document points out: "the Islamic State cannot renew the *Ummah*'s blood without a human member capable of always producing." It includes, among the stated aims of its internal societal program, "[r]aising a knowledgeable Islamic generation capable of bearing the *Ummah* and its future without needing the expertises of the West."

ISIL's propagandistic impulses extend to what seem to be the administrative details of governance; it appears to view even devising its own calendar and redrawing its internal borders as a way to counter its enemies' information operations campaigns. The fifth chapter of the "Principles" document describes the previous drawing of borders after Sykes-Picot as a "dirty political decision" that drew "social, *madhhabist* [doctrinal], and ethnic differences in every region," thereby "deepening the roots of the differences between the Sunni Muslims." Likewise, in late 2014, the ISIL Shari'a Committee for the Observation of New Moons announced that it would work on devising a new calendar "until the Islamic State should be independent in its calendar and observation for the new moons from all the states of *tawagheet* [idolatrous tyranny] who have subsumed this issue for their personal and political whims and aims" (Archive of Islamic State Administrative Documents, Specimen 2P).

The third major prong of ISIL's information operations is its media and propaganda system. The "Principles" document explains that, in the interest of advancing the aims of territorial control, resource control, and internal and external recruitment, "all of [the state's] ideas and activities should be advertised." To this end, the document instructs that one media foundation be in charge of "supervising the distribution of the media offices in the provinces and the [other] media foundations," defining publication and broadcasting priorities, inspecting the other foundations, and training media staff. The document tasks the provincial media offices with "covering the military operations and their results . . . as well as services' facilities, implementing *Shari'a* rulings and the course of life in the province," and printing and distributing these media productions within the province. The document stipulates that the other, auxiliary media foundations, should "not . . . be allowed to cover security operations or implementations of [judicial] rulings."

ISIL's propaganda system also serves as the connective tissue that allows ISIL's deliberate embrace of atrocities to intimidate potential opponents and to lure its more bloody-minded potential supporters to its cause. This concept of the theater of violence is laid out in Naji's foundational document, *The Management of Savagery*. Will McCants, of the Brookings Foundation, argues that this strategy manual, which evaluates tactics according to their psychological effects both on enemies and potential supporters, is "revered" by ISIL operatives (2015). As the means that allows ISIL's recruitment and governance efforts to be globally seen and known, ISIL's propaganda system thus enables and multiplies the effects of the other two main prongs of ISIL's information operations campaign.

Economic Subversion and Control

The "Principles" document characterizes the controlling of economic resources—and specifically establishing independent sources of money, weapons, oil, gas, and food—as a key element of ISIL's attempt to gain leverage against its opponents. It argues that ISIL must establish "secure financial resources," including "oil and gas and what the land possesses including gold as currency that does not deteriorate or decline," so that its enemies "cannot pretend that it [the Islamic State] has no existence and might." The chapter on the administration of wealth contends that "the might of the Islamic State can only be through its being free entirely from all bonds of tyranny that the West possesses as means of leverage against it." These means of leverage include the "need of the *mujahideen* for support, wealth and weapons." ISIL seeks resource independence in order to deny leverage to its enemies.

Accordingly, the plan laid out in "Principles" for controlling resources includes overseeing the production of all oil and gas extraction, manufacturing, and production; requiring specific state authorization for foreign investment in, extraction of, and production of oil and gas, gold and antiquities, weapons and ammunition, and water, flour, and livestock; setting up "factories for local military and food production" in order to attain "independence from the monopoly of arms dealers"; ensuring that each province must reduce excess expenditures, as each province "must operate independently"; establishing trade routes internal to the state; and conducting trade outside the state "without an intermediary" in order to avoid a "monopoly of the intermediary for business transactions and means of connection." These are all forms of economic control that aim to establish independence in order to reduce adversaries' leverage over ISIL.

This objective of resource independence appears to extend to currency itself. In the November 2014 issue of *Dabiq*, ISIL introduces its decision to mint its own currency as "an effort to disentangle the *Ummah* [Muslim community] from the corrupt, interest-based global financial system." ISIL claims, in its hour-long video "Return of the Gold Dinar," that it decided to mint its own currency in order to "break the U.S. capitalist financial system of enslavement" (Gidda, 2015). Even if the effect so far has been merely symbolic, ISIL has presented its decision to mint its own currency as an attempt to reduce its dependence on external financial systems and undermine the global economy by hurting the U.S. dollar and global financial system (Gidda, 2015).

Diplomacy

While the extent to which ISIL views enduring agreements with foreign nations permissible remains ambiguous, foundational documents do acknowledge the potential value of temporary treaties.

ISIL's broad perspective on foreign relations appears to be that relations with other states or organizations may be entered into only if they are a valuable source of leverage. The "Principles" document instructs that "external relations . . . are among the foundations that show the strength and might of the state [and] alliances should be . . . a guarantee of force and leverage that the Islamic leadership can use in all its matters with the external world." The document specifies that "leadership is not allowed to . . . ally with a state or implement an agreement with it" if the agreement "violates sharia politics." It mandates that any agreement must not include any provisions, even future ones, which "[touch] on the freedom and sovereignty of the Muslim state," such as "bonds of debt or conditions of harmful exploitation."

ISIL's foreign relations doctrine apparently prohibits making treaties with certain nations. The "Principles" document stipulates that ISIL "is not allowed to deal with another state that has a history of hostility to Islam's spread, the building of mosques and oppression of Muslims in its land." ISIL would almost certainly apply this prohibition on diplomacy to countries such as the United States.

An article in *Dabiq* appears to complicate ISIL's treatment of this issue, suggesting that temporary treaties with even Western nations may be possible. An article attributed to John Cantile in the March 2015 issue of the magazine argues that the United States and its Western allies should consider a truce with ISIL. Attributing this article to John Cantile, ISIL's British hostage, allows ISIL to distance itself from the proposal. However, the fact that ISIL included the article at all is suggestive. A footnoted "Editor's Note," representing the ISIL establishment's response to the idea, cautions readers: "A halt of war between the Muslims and the *kuffār* [infidels] can never be permanent, as war against the *kuffār* is the default obligation upon the Muslims only to be temporarily halted by truce for a greater *shar'ī* interest," citing the precedent of the Prophet's ten-year treaty with the Hudayba in Mecca. Even the text attributed to John Cantile includes the following caveat: "Any truce between the West and the Islamic State would ultimately have to address the end of support for Arab and non-Arab tyrannical puppets in the Muslim world as well as an end of support for Israel." While the idea of a truce between ISIL and Western nations is unrealistic, the possibility of ISIL making an at least temporary agreement with some more flexible government appears to exist.

ISIL's Application of Political Warfare Tactics

Diplomacy

ISIL's use of diplomacy so far seems to be restricted to nonstate actors, with the possible exception of tacit coordination with the Syrian regime. Its use of affiliates involves proxies formally incorporated as franchised "provinces." Its postures toward rival groups vary, from friendly overtures to (or informal alliances with) independent militant groups, such as Ansar al-Sharia, to open enmity and attempts to discredit Jabhat al-Nusra and al Qaeda in the Arabian Peninsula and poach their supporters.

Projecting an image of global expansion, ISIL has collected pledges of loyalty from a range of violent Islamic extremist organizations and has officially incorporated some of them into the system of ISIL "provinces" (*wilayat*). ISIL has currently declared as many as 40 provinces: 12 in Iraq, 11 in Syria, 6 in Yemen, 3 in Libya, 2 in Saudi Arabia, 2 in Afghanistan, and 4 other provinces known as the Algerian Province, West African Province, Qawqaz Province in the Caucasus, and Sinai Province in Egypt (Winter, 2015; Zelin, "Picture or It Didn't Happen," 2015).

As an example of this incorporation process, the Egyptian terror group Ansar Beit al-Maqdis (ABM) pledged allegiance to ISIL in November 2014; the pledge of allegiance was accepted, and ABM was accordingly renamed "Sinai Province" (Kirkpatrick, 2014). ABM's declared aims had included destroying Israel, implementing sharia law, and establishing an Islamic caliphate in the Sinai ("Ansar Bayt Al-Maqdis, 2015"). ABM began its militant activities in 2011 by launching rocket attacks against Israel but shifted to targeting Egyptian security and police forces in 2013 (Kingsley, 2015), avowedly in response to both the security force's crackdown on Islamist dissidents after Egyptian president Muhammad Morsi was removed from power and the state's ongoing counterinsurgency in the Sinai Peninsula (Barnett, 2013). Now, in its role as the Sinai Province of the Islamic State, ABM seeks to co-opt marginalized populations in the Sinai Peninsula by handing out flour and cash (El-Ghobashy, 2015). These attempts are a small-scale imitation of ISIL's efforts at governance structures in Syria and Iraq.

As an example of a failed attempt at using economic inducements to attract affiliates, ISIL allegedly tried to enlist the Pakistani group Ansar al-Furqan to pledge allegiance to ISIL and join the organization as its "Baluchistan Province." According to an individual claiming to belong to Ansar al-Furqan, after the group declined, ISIL attempted to "buy" its allegiance (SITE Intelligence Group, 2015).

While ISIL's diplomatic and political interactions are largely aimed at affiliates and fellow adversary groups, it has some dealings with states. Despite its ideological rigidity, ISIL has demonstrated pragmatism as needed to achieve its ends or ensure its survival.

While hostage negotiations can hardly be considered meaningful diplomatic engagements, it is worth mentioning that ISIL has released some prisoners according to agreements reached with opposing governments. While the U.S. and British governments prohibited families of ISIL hostages from paying ransom to recover them, ISIL hostage Rye Ottoman was released in June 2015 after his family paid ISIL millions of dollars as a ransom (Orange, 2015). On a more government-to-government level, in September 2014, Turkish prime minister Ahmet Davutoglu announced that, following "intense efforts that lasted days and weeks," ISIL had handed over 46 Turkish hostages to the Turkish intelligence agency. Similarly, in June of that year, 32 Turkish truck drivers whom ISIL had seized were released within a month, although the Turkish government provided no information regarding their release ("ISIL Released Turks Held Hostage in Iraq," Al-Jazeera, 2014).

More significantly, ISIL appears to have coordinated at least tacitly with the Assad regime. In the wake of the U.S. invasion of Iraq in 2003, ISIL's precursor AQI apparently enjoyed special status with the Assad regime, which reportedly facilitated a massive flow of Islamic militants into Iraq (Scarborough, 2013). Iraqi officials reported that Syrian border guards were assisting Islamic militants in openly crossing the border into Iraq; many of these militants joined the AQI-affiliated insurgency against the U.S. occupation. Later, in the early days of the Arab Spring, the Assad regime reportedly released Islamic radicals from its prisons (Engel, 2014). As recently as January 2014, during fighting between ISIL and other rebel factions, Der Spiegel reported that Syrian regime aircraft "regularly bombed only rebel positions, while the Islamic State emir ordered his fighters to refrain from shooting at the army" (Reuter, "The Terror Strategist," 2015). This cooperation also seems to have extended to the economic front via oil smuggling.

Economic Subversion and Control

ISIL's economic tactics include oil smuggling, resource control, extortion, and minting its own currency. As ISIL's internal "Principles" strategy document argues, establishing independent sources of oil, gas, food, revenue, and currency is integral to its strategy of reducing its opponents' leverage as well as establishing its legitimacy as a state in order to reap the recruitment benefits of statehood. ISIL seeks resource independence in order to strengthen its claim to statehood and to deny leverage to its enemies.

ISIL's tacit cooperation with the Assad regime appears to extend to smuggling oil to the Syrian government as well as to some of ISIL's other publicly avowed enemies. Crude oil travels via middlemen to the Assad regime forces, Kurdish areas, Turkey, and even to Syrian rebels supported by the United States, according to RAND expert Ben Bahney (Behn, 2015). In November, the U.S. government levied sanctions on George Aswani, who had allegedly been serving as an intermediary for the oil trade between ISIL and the Assad regime ("Civilians Pay," IRIN News, 2015). While Kurdish parliament members have denied accounts of ISIL selling oil in Iraqi Kurdish areas, KRG officials had made 15 arrests of businessmen and military officials suspected of trading with ISIL as of October 2015 (Hendawi and Abdul-Zahra, 2015).

ISIL's export of oil to Turkey garnered significant negative attention from Western governments, and Turkey responded with steps to dampen the smuggling operations and regain control of its side of the 100-kilometer length of its border through which foreign fighters were entering ISIL-held territory and ISIL crude oil was exiting to Turkey (Behn, 2015). There have also been reports of equipment for extraction and refinement being imported across the Turkish border into ISIL-held territory. In late 2015 the Turkish prime minister's office claimed that it had "effectively stopped oil smuggling" from ISIL-held territory, claiming that it had thwarted 3,319 acts of smuggling into Turkey from Syria and apprehended over 5.5 million liters of smuggled oil (Hendawi and Abdul-Zahra, 2015). RAND expert Howard Shatz noted in November 2015 (Interview with Howard Shatz, 2015) that Turkey had launched a major operation to stop the flow of ISIL oil to Turkey via pipeline (Ganor, 2015).

ISIL appears to have partially achieved its objectives of controlling its own critical resources and sources of revenue, fuel, and food in a bid for independence from foreign funders. As of December 2015, approximately half of its monthly revenues, assessed at around $80 million, were generated by taxes and property confiscation; the other half were generated primarily by black market oil sales inside and outside of ISIL-held territory (Kaplan, 2015).

ISIL closely tracks its sophisticated oil and gas production and distribution system. Oil and Gas Divisions generate receipts for oil and gas sales, tracking and regulating the flow of fuel inside and outside its territory (Archive of Islamic State Administrative Documents, Specimen 5U). The Rikaz Department in Al-Khair Province sent out a notice outlining the procedures for distribution of "gas bottles" to the populace; residents must present a "marriage proof document" such as a "family register" or "original marriage bond" in order to receive the fuel (Archive of Islamic State Administrative Documents, Specimen 4R).

ISIL has also sought to control sources of food within the territory it occupies. An ISIL report obtained by *Der Spiegel* in 2014 listed pretexts, ranging from alleged embezzlement to dissolute behavior, that ISIL could use to seize a flour mill as part of a veiled campaign to seize "all strategically important facilities like industrial bakeries, grain silos and generators" (Reuter, "The Terror Strategist," 2015). When sweeping across Iraq in 2014, ISIL reportedly seized wheat stored in silos, in amounts worth up to $200 million. ISIL now controls an expanse of fields that, as Bloomberg's Cam Simpson and Matthew Phillips reported in November 2015, have "historically produced half of Syria's annual wheat crop, about one-third of Iraq's, and almost 40 percent of Iraqi barley, according to UN agricultural officials and a Syrian economist" (2015).

ISIL's objective of reducing its dependence on external resources in order to weaken its opponents' leverage also extends to currency. ISIL has substantial currency reserves, partially generated by one-time acquisitions, such as when it looted Mosul's banks, which alone may have accounted for between $500 million and $1 billion (Trumbull, 2015).

ISIL has begun minting its own currency, vowing to return to the use of precious metals as the basis for value and as an attempt to reduce its dependence on external financial systems and undermine the global economy. However, these moves may be mostly symbolic; a month after the "Return of the Gold Dinar" video in August 2015 denounced the "U.S. capitalist financial system of enslavement," ISIL was still paying its fighters in U.S. dollars (Gidda, 2015).

On other occasions, ISIL has sought to undercut the use of other nations' currencies directly. In September 2015, the Hisba Department of Aleppo Province prohibited the use of two new Syrian banknotes, enjoining "all exchange shops" that they "must not exchange or circulate these notes. Rather they are to be received from people and removed from the lands of the Islamic State" (Jawwad, 2015).

Military and Intelligence

ISIL's military tactics have been covered extensively elsewhere, so this section will focus on ISIL's intelligence and counterintelligence methods. However, it is still worth noting ISIL's exceptionally heavy and deliberate use of irregular warfare tactics and barbaric killings as part of a psychological operations campaign. ISIL repeatedly surpasses its parent organization, al Qaeda, in terms of its well-publicized atrocities, which are both mass-produced and carefully orchestrated for maximum shock value.

ISIL's brutal campaign against the anti-ISIL vigilante group Mosul Brigades offers a prime example of ISIL's exploitation of the theater of violence. In June 2015, ISIL released a video showing 16 men, allegedly including some members of the Mosul Brigades, being blown up in a car, drowned in a cage, and decapitated by explosive neck-collars, in a naked attempt at intimidating the Mosul-based resistance (Lando, 2015). ISIL in Mosul has undertaken a campaign of systematically killing Mosul Brigade members. ISIL also retaliates against assassinations of its own militants by accusing former ISIL security personnel of espionage and collaboration with the Iraqi army and publicly executing them in order to intimidate others (Paraszczuk, 2015).

The litany of ISIL's terrorizing atrocities is near endless. In the latter half of 2014, ISIL massacred more than 500 members of the Albu Nimr tribe, which had resisted ISIL's advance, burying 200 Albu Nimr bodies in a single ditch near Hit (Wolf, 2015). Over the course of three days in Sirte, Libya, ISIL militants killed over 50 members of the Feryan tribe, which had refused to pledge allegiance to the organization ("Libya," 2015). ISIL has undertaken a campaign of sexual slavery against the Yazidi in Iraq; as a female author wrote in the May 2015 issue of ISIL's *Dabiq* magazine, "I and those with me at home prostrated to Allah in gratitude on the day the first slave-girl entered our home. Yes, we thanked our Lord for having let us live to the day we saw *kufr* [infidels] humiliated and its banner destroyed." A captured head logistician for ISIL suicide attacks in Baghdad said, when asked about how he selected his targets: "It was about hitting as many people as possible—especially police officers, soldiers and Shiites. . . . I thought, at some point these Shiites . . . who experienced an explosion would start to think and that they would be afraid. . . . My idea was to continue until all of them converted. Or emigrated" (Reuter, "'I'm Not a Butcher,'" 2015).

But the focus here is on how ISIL has employed intelligence and counterintelligence measures to great effect. Laying the groundwork for its rapid expansion, ISIL in 2012 and 2013 began infiltrating villages under the pretext of setting up Islamic missionary centers, selecting a few attendees to spy on their own villages. These attendees were asked to assemble lists of powerful families, their sources of income, local brigades and their leaders' political orientations, and information usable for blackmail, such as homosexuality or extramarital affairs. After using this covert infiltration and intelligence-gathering to prepare the ground for takeover, ISIL would then move into the villages openly, systematically eliminating potential opponents. In Raqqa in the summer of 2013, disappearances quickly climbed into the hundreds.

Parallel structures of ISIL security emirs are replicated at the local, district, and provincial levels of administration; the ISIL intelligence services monitor not only external enemies and local populaces, but also ISIL fighters and officials (Reuter, "The Terror Strategist," 2015). A trove of ISIL intelligence documents obtained by Der Spiegel in 2014 revealed an extensive surveillance system, including lists of Assad regime spies who had been placed in rebel factions and lists of ISIL's own informants who had been placed in both rebel factions and government militias (Reuter, "The Terror Strategist," 2015). ISIL has also infiltrated Kurdish militias (Reuter, "Rational Monster," 2015).

ISIL's current intelligence branches continue to gather information on internal threats, especially tribes. In March 2015, when sending a statement to "the sheikhs and tribes of Wilayat Ninawa" requiring their presence at a conference, the Tribes Department in Ninawa Province copied the Islamic Military Police, Islamic Police, and ISIL intelligence services on its request to "let us know immediately those who reject and refuse" (Archive of Islamic State Administrative Documents, Specimen 4I). An ISIL administrative form from the Committee to Track the Affairs of the Tribes has various columns for noting each individual's age, position, and place of residence, in addition to whether or not he had participated in fighting ISIL and whether or not he had handed over his weapons (Archive of Islamic State Administrative Documents, Specimen 8C).

For ISIL's counterintelligence operatives, the Internet is an area of great concern. In March 2015, the Aleppo Province general administration reminded its "brothers not to disclose news and information of the movement of [their] brothers through use of the Internet," counseling them: "When the Prophet (PBUH) wanted to conduct a raid, he would conceal it under another purpose. . . . And he would say: 'War is deceit'" (Archive of Islamic State Administrative Documents, Specimen 5E). An Intelligence Services Official in Furat Province sent a "Central Public Security Directorate" to be distributed to "all Internet users"; it forbade the use of the Internet, warning, "Whoever is found with an Internet connection inside his home, office or any private place will expose himself to severe reckoning" (Archive of Islamic State Administrative Documents, Specimen 9A). Other provincial agencies have taken different measures, such as the Public Security Department of Raqqa Province, which announced in July 2015, "To all Internet shop owners. . . . You must ensure to record the IDs of users except soldiers of the Islamic State and their families" (Archive of Islamic State Administrative Documents, Specimen 6O). Shortly afterward, the province restricted Internet use to the registered Internet cafés, thereby tracking all Internet users other than ISIL soldiers and family members (Archive of Islamic State Administrative Documents, Specimen 9A).

These counterintelligence measures extend to controlling mobile phone usage. In 2014, the Ninawa Province Media Office announced that mobile phone networks would be cut off in Mosul, after spies inside the province had been "confirmed" to be using "various devices of connection to provide the enemies of the Islamic State with information that has brought about losses to the Dawla of many of its knights" and after discovering that anti-ISIL coalition "aircraft ha[d] been able to monitor the phones of the mujahideen and follow their movements" (Archive of Islamic State Administrative Documents, Specimen 8Z).

Other counterintelligence measures may include fighters wearing black masks, not only to protect identity but also to make it more difficult for enemies to differentiate units and assemble estimates of forces. At other times, manipulating expectations of a black wardrobe, groups of ISIL fighters have been known to infiltrate disorganized associations of Syrian rebel factions by simply changing into jeans and vests (Reuter, "The Terror Strategist," 2015). As a countermeasure against the Mosul Brigades—who systematically target and assassinate ISIL militants using strangling, sniper rifles, and booby-trapped vehicles—ISIL has issued guidelines advising its fighters in Mosul to wear black masks to avoid being identified, to drive unmarked cars, and to wear beards to blend in better. ISIL has also reportedly tried to conceal the Mosul Brigade's successful assassinations, explaining away the deaths of fighters strangled, stabbed, or killed by guns with silencers as the result of being killed in combat or by coalition air strikes, and bringing their bodies to and from the morgue under cover of darkness (Paraszczuk, 2015).

In such a way, ISIL's sophisticated intelligence and counterintelligence apparatus weaves seamlessly with its deliberate embrace of atrocities in order to achieve maximum psychological impact. This is achieved, of course, with the assistance of its information operations campaign.

Informational and Cyber

ISIL employs a multilayered propaganda production system and a resilient, decentralized dissemination method to great effect in its efforts to recruit supporters and pressure its opponents. Foreign fighters have answered Abu Bakr al-Baghdadi's call for *hijra* with an alarming surge of emigration, spurred on in part by ISIL information operations on social media and targeted recruitment campaigns, which are facilitated by a radicalization process that involves a shift from open to encrypted communications. Cyber-warfare, or at least publicly observable cyber-warfare, appears to be far less central to the ISIL information operations effort than media productions or recruitment.

ISIL exploits social media in innovative ways. Partially decentralizing its propaganda production and almost entirely decentralizing its propaganda dissemination, ISIL relies on a multilayered set of media centers to produce its official media publications, then leverages an extensive and relentless network of supporters on social media in order to distribute its message.

A range of outlets produces official ISIL propaganda. Al-Furqan Media, Al-Hayat Media, Ajnad Media, and Al-I'tisam Media produce and release top-level messaging, such as the sermons of Abu Bakr al-Baghdadi, audio addresses from official ISIL spokesperson Abu Muhammad al-Adnani, and issues of the sleek English-language magazine *Dabiq*, French-language magazine *Dar al-Islam*, and Turkish-language magazine *Constantinople*. Other outlets include the A'maq News Agency and Al-Bayan Radio Station (Zelin, "Picture or It Didn't Happen"). Beyond those centrally controlled media organizations, ISIL provincial media offices produce locally focused situation reports and messages, such as the "Victory from God and an Imminent Conquest" video series from the Al-Khayr Province Media Office, or "To the Defiant Tribes of the Sinai," a statement from the Sinai Province Media Office (Zelin, "New Video Message from the Islamic State"; "New Statement from the Islamic State").

Aaron Zelin's analysis of the 123 media releases from ISIL in a single week in April 2015 found that top-level ISIL outlets represented 22 percent of the media releases, while the other 78 percent came from province-level media centers, suggesting that ISIL media production has decentralized drastically over the last two years. Zelin, who runs Jihadology, the jihadist primary source material online archive, also found that while only 58 percent of ISIL's provinces are in Iraq and Syria, a disproportionate number (81 percent) of the media releases came from provinces inside the two countries; Libyan provinces accounted for a distant third, at 11 percent of the 123 releases in the study (Zelin, "Picture or It Didn't Happen").

Zelin also argues that over the latter half of 2015, the ISIL media machine's output had declined in terms of both quality and quantity. Zelin tracked the number of nonmilitary ISIL photos from Syria and Iraq from January to November 2015. For each country, counted separately on a three-month rolling basis, he found a steady rise from January until the peak months of June and August 2015, followed by a steady decline until the data ended at the end of November 2015. He attributed the decline to the elimination of ISIL media operatives and territory lost by ISIL, and cited experts such as Cori Dauber and J. M. Berger as agreeing with him that the quality of videos from provincial media offices had also declined, with the exception of those released by Raqqa Province (Zelin and Fellow, 2015).

In order to disseminate its official propaganda, ISIL has shifted away from a broadcast model, characterized by one-to-many sharing, to a dispersed and resilient form that relies on peer-to-peer sharing and deliberate lateral redundancy across multiple platforms. ISIL initially relied on official Twitter accounts to release its productions; after Twitter began suspending pro-ISIL accounts, ISIL adapted. It adopted what Ali Fisher, in his study on jihadi proliferation of social media, calls a user-curated "Swarmcast" method (2015). By crowd-sourcing the hosting and popularization of its content, ISIL inoculated its information operations from attempts to remove and block its content.

As soon as Twitter suspends one ISIL supporter's account, the supporter creates a new account or moves to his previously created backup account, the name of which he has already promulgated to other supporters (SITE Intel Group, "Jihadists Circulate Guide," 2015). As one ISIL supporter urged in an online blog, "To counter suspensions, you must support each other. Don't make the brothers and sisters ask for a shout out. If someone you recognize has followed you after a suspension, tell your followers to follow them." YouTube videos are downloaded and re-uploaded by as many users as possible in order to achieve viral growth and maintain access to the content (SITE Intel Group, "Blog Post Suggests Ways," 2015). Instead of following an official ISIL spokesperson account, which can easily be suspended, pro-ISIL users read and retweet posts on pro-ISIL hashtag channels. Even when hosting sites, such as YouTube and JustPaste.It, comply with requests to take down ISIL propaganda, supporters who downloaded the content before it could be removed simply re-upload the content and share the new link on a pro-ISIL Twitter hashtag (Fishwick, 2014).

ISIL content hosts and promoters are proactive and myriad, while the effort to block ISIL content and suspend ISIL users is reactive, with only a finite number of people empowered and dedicated to carrying it out. For instance, by the time the U.S. State Department counter-ISIL Twitter account (*@dsdotar*) returned from a weekend hiatus on Monday, May 19, 2015, the newly released fourth installment of the "Clanging of the Swords" video series had been viewed over 100,000 times, and enough users had downloaded the content to make it effectively permanently available online (Fisher). Even though the Brookings study on ISIL supporters online argued that the rate of Twitter suspensions may have limited the ISIL network's growth, it also acknowledged that suspensions cannot be expected to fully interdict the spread of ISIL propaganda on Twitter (Berger and Morgan, 2015).

ISIL also welcomes supporters and members taking the initiative to create their own unofficial ISIL propaganda, in addition to disseminating official content (Winter, 2015). In Klausen's 2014 study of tweets posted by 29,000 Twitter accounts belonging to Western ISIL foreign fighters, she found that disseminators outside the conflict zone, including a handful of particularly influential women, played a key role in this effort to "build redundancy by spreading the material, often posting and reposting material provided by the feeder accounts belonging to organizations and fighters based in Syria." A report by the International Centre for the Study of Radicalisation and Political Violence (ICSR) also emphasized the role of disseminators based primarily in the West, arguing that many foreign fighters informed themselves about the conflict by following unofficial disseminators rather than official ISIL channels (Carter, Maher, and Newman, 2014).

However, Klausen argues that while self-publication sites like Twitter and YouTube give online information the appearance of spontaneity and authenticity, the high degree of coordination between posters was so high that it was "improbable that it could have occurred by chance." She found that ISIL closely regulates which of its fighters provide updates from the battlefield, as new arrivals to training camps must surrender their cell phones and "[u]nauthorized contact with family members" was allegedly punishable by death. The most frequently tweeted content in the dataset consisted of either religious instruction or battlefield reports; less than a fifth were interpersonal communications, and only a handful of tweets consisted of threats against the West, suggesting substantial ideological conformity across posts.

The effectiveness of ISIL's social media campaigns can be hard to measure, but ISIL has clearly been highly successful in attracting foreign recruits. *Der Spiegel* reports that in 2012 ISIL leadership opted to focus on recruiting foreigners rather than local Syrians, and initially training them separately; the lack of connection to the populace made them more reliable and merciless enforcers, as well as easier to quickly move from place to place (Reuter, "The Terror Strategist," 2015).

This effort has expanded drastically in the years since, particularly after ISIL's dramatic territorial gains and declaration of itself as a caliphate. As Nicholas Rasmussen, director of the National Counterterrorism Center, testified to Congress in February 2015, "The rate of foreign fighter travel to Syria is unprecedented. It exceeds the rate of travelers who went to Afghanistan and Pakistan, Iraq, Yemen, or Somalia at any point in the last 20 years." The Soufan Group estimated that between June 2014 and December 2015, the number of foreign fighters flowing into Syria had more than doubled, and concurred with the May 2015 assessment of the UN Security Council monitoring team that the number of foreign fighters who had traveled to join the current conflict in Iraq and Syria may have been as high as 30,000, from over 100 countries. While over a third of the fighters originate from Western countries, including the European Union, Canada, the United States, New Zealand, Australia, Russia, and the Balkans, slightly over half come from the Arab world, particularly Tunisia, Saudi Arabia, Morocco, Jordan, and Turkey (Barrett, 2015).

By appealing to targets' disaffection and lack of belonging, as well as to a desire to help protect a Muslim community under attack, ISIL's use of social media sites such as Twitter and Facebook has proven highly effective at spreading its ideology and plays a significant role in its recruitment and fund-raising efforts (Barrett, 2014). While exposure to propaganda alone does not turn opponents or even potential recruits into members, it can furnish dissatisfied individuals with an internally coherent system of beliefs and significantly catalyze the radicalization process (Winter, 2015).

This radicalization process is assisted by ISIL's online targeted recruitment activities. According to Henry Tuck from the Institute for Strategic Dialogue, ISIL has "units of specialized recruiters operating around the clock from Internet cafes in Iraq and Syria, interacting on an individual level with prospective recruits" (Shroder, 2015). J. M. Berger adds that while in-person recruitment has also been observed, it is dwarfed by online recruitment in countries such as the United States, and offers this rule of thumb: "In general, the role of offline recruitment becomes more significant the closer one gets to an Islamic State–contested territory (2015).

Berger's 2015 study on ISIL's use of online tailored interventions to turn passive supporters into active contributors suggests a common set of techniques used by recruiters both official and unofficial. After first contact is established, whether sought by the recruiter or the target, the recruiter constructs a "micro-community" around the target, keeping up an almost constant stream of contact and encouraging the target to "insulate against outside influences." This is followed by a key turning point: a "shift to private communications," such as direct messaging on Twitter or Facebook, or messaging applications like WhatsApp and Kik. At this point, the recruiter identifies and encourages the target to take action on behalf of ISIL, whether through social media activism, making *hijra* and emigrating to ISIL-held territory, or conducting terrorist attacks in the target's home country.

The step in this radicalization process of urging a target to insulate himself from opposing influences is mirrored in a March 2015 technical guide from Ansaru Khilafa, a pro-ISIL website, which advises readers on how to subscribe to "the blocklist" on Twitter: a method of automatically blocking an ever-growing swath of anti-ISIL accounts. The author explains, "Everytime the owner of the blocklist blocks someone, your account will automatically block the account as well." The guide encourages readers to block even Muslims, if those Muslims had been following the "major spammers which are taking down" pro-ISIL accounts or were somehow affiliated with enemies of ISIL.

Tech-savvy ISIL supporters also share advanced techniques intended to guard against detection. Another guide posted by the Ansaru Khilafa technical team gives detailed, step-by-step instructions to ISIL supporters on how to use GPG4USB to encrypt files and text messages, how to encrypt the messages while offline, and how to send the encrypted messages via Tor-based e-mail channels ("Registering on Twitter Without Phone Verification, Part 1"; "Registering on Twitter Without Phone Verification, Part 2"). The previous month, the team posted a sophisticated two-part guide advising ISIL supporters on how to evade Twitter's phone verification requirement when registering for an account from an IP address outside of the Untied States. The guide's suggested methods include using a proxy VPN such as CyberGhost in order to register with Twitter using a U.S.-based IP address, or using JonDo IP-Changer proxy software.[1] The guide reminds ISIL supporters not to use an e-mail address for which they used their real IP address to register, and advises that they avoid GMail, Microsoft, and Yahoo! and instead register for Tor-based e-mails such as Mail2Tor, Sigaint, or RuggedInbox.com while using the Tor Browser Bundle.[2] The guide cautions that "this method does not provide 100% anonymity if the intel agencies truly consider you a high-value target," warning that intelligence agencies will ask Twitter and then CyberGhost for the IP addresses used to register for their services and accordingly be able

1 http://www.cyberghostvpn.com/; https://anonymous-proxy-servers.net/en/jondo.html

2 http://mail2tor2zyjdctd.onion; http://sigaintevyh2rzvw.onion/; https://ruggedinbox.com/; https://www.torproject.org/projects/torbrowser.html.en

to identify the user's real IP address. The guide, citing a range of NSA capabilities, further counsels that intelligence agencies can decrypt and hijack VPNs for anyone who is actually under surveillance, but says, "This danger does not apply to registering with the e-mail services I mentioned as you registered with them through Tor, which as of our current understanding, is secure."

The movement of some ISIL supporters from open to encrypted communications is sometimes mirrored at the level of ISIL media organizations. Two days after the Paris attacks in November 2015, an ISIL propaganda dissemination telegram channel announced that it was launching a "dark web" version of the Isdarat website, which archives ISIL propaganda, and provided links using .onion for both Tor and normal users (SITE Intelligence Group).[3] FBI director James Comey testified to Congress in July 2015 about the threat posed by ISIL supporters' shift to encrypted direct messaging platforms.

Cyber warfare, or at least publicly observable cyber warfare, appears to be far less central to the ISIL information operations effort than media productions or recruitment. Publicly visible cyber offensives, led by self-proclaimed, pro-ISIL hacking groups such as the Cyber Caliphate and Islamic Cyber Army (ICA), appear to consist primarily of hacking websites and either defacing them or reposting personal information about ISIL's enemies, which sometimes had already been available publicly online. These releases often consist of "kill lists" of U.S. military personnel's location and contact information (SITE Intelligence Group, 2015). For instance, in November 2015, the Cyber Caliphate posted names, phone numbers, zip codes, and departments of 36 U.S. military personnel, sourcing the information to a Justpaste.it document created in August 2015 (SITE Intelligence Group).

A particularly active pro-ISIL hacking group that was formed in early September 2015, the ICA undertook seven hacking campaigns in its first two months; however, SITE Intelligence Group assesses that the ICA's capabilities have not been shown to "extend beyond basic defacements and re-posting data from publicly-available sources

---

[3] For Tor users: http://isdratetp4donyfy.onion; for non-Tor users: http://isdratetp4donyfy.onion.link

or previously-claimed alleged hacks." Sometimes, the claims of hacking are clearly disingenuous. For instance, when the ICA disseminated physical and e-mail addresses of French government officials, including the address of France's Élysée Palace, on Twitter and Telegram and claimed that this data had come from hacking operations under its "#WorldUnderHacks" campaign, the information had already been publicly available. However, not all of the ICA's claims can be so handily dismissed as puffery; under the same campaign, the ICA posted 276,817 Twitter usernames and passwords, which they purported to have hacked, when the account information had not already been available online (SITE Intelligence, 2015).

With the exception of hacks of personal information used to populate public "kill lists," pro-ISIL hackers appear to focus on website defacement, which is, in essence, another form of propaganda intended to intimidate opponents and galvanize supporters. Website defacements and "kill lists" are sometimes linked. On another occasion, the ICA defaced Iranian websites with a message threatening to publish "sensitive and classified information" of the Iranian government, including personal information of Iranian military personnel, if Iran did not stop its "war on the Islamic State." The ICA claimed to have taken data from the Iranian Ministry of Health and the Ministry of Education (SITE Intelligence, 2015). Another pro-ISIL hacking group, calling itself Rabita Al-Ansar, attacked the Bermuda Broadcasting Company website in May 2015 and displayed photographs that purportedly showed the television programming having been changed to pro-ISIL propaganda.

Another facet of ISIL's information operations is its campaign of sexual slavery. ISIL's systematic campaign of enslavement and sexual assault, particularly against the Yazidi women of Iraq, can be seen as a means of intimidation but also as a recruitment tool. The October 2014 *Dabiq* article "Revival of Slavery before the Hour," in which ISIL declares its embrace of the abhorrent practice, claims that "the desertion of slavery had led to an increase in *fahishah* (adultery, fornication, etc.), because the *shar'i* [legitimate] alternative to marriage is not available, so a man who cannot afford marriage to a free woman finds himself surrounded by temptation towards sin." When coupled with ISIL's promises to match up emigrant husbands and wives, this may help lure men who are excited by this twisted version of religiously sanctioned sexual "freedom" or simply by the idea of sexual violence.

To further support the practice of sexual slavery, ISIL commissioned a woman named "Umm Sumayyah Al-Muuhajirah" to defend the practice. In an article entitled "Slave Girls or Prostitutes?" in the May 2015 edition of *Dabiq*, she wrote that she was alarmed that Islamic State supporters denied the use of female slaves "as if the soldiers of the Khilāfah had committed a mistake or evil." She continued to defend the practice by asserting it is not intended for male fighters' sexual pleasure nor the destruction of an enemy, but as a traditional means to encourage the slaves to convert to Islam. She proclaimed,

> I write this while the letters drip of pride. Yes, O religions of *kufr* [disbelief] altogether, we have indeed raided and captured the *kāfirah* [infidel] women, and drove them like sheep by the edge of the sword. . . . Or did you and your supporters think we were joking on the day we announced the *Khilāfah* [caliphate] upon the prophetic methodology?

By choosing a woman to defend sexual slavery, the editors of *Dabiq* carefully shaped their message in an attempt to appeal to previous or potential supporters who had denounced the practice as an affront to women.

ISIL's multilayered information campaign is not only relentless, using a decentralized dissemination method to amplify its resilience and reach, but sophisticated and deliberate, with targeted recruiting and a multilingual message carefully shaped to appeal to different audiences.

Examining How ISIL Messaging Varies Across Language Platforms

In this section we present original research findings on how ISIL's English-language propaganda quantitatively differs from its Arabic-language propaganda, as well as the implications of those differences. This effort applies a lexical analysis method to quantitatively compare

English and Arabic text corpora, constructed from a year of ISIL publications and releases in both English and Arabic. The analysis finds that while ISIL's Arabic messaging, apparently targeted more to people already within the rank and file, emphasizes violence and violent jihad, ISIL's English messaging has a more international focus.

Previous Research

Observers agree that ISIL targets its messages to a variety of linguistic audiences. While Arabic dominates the ISIL propaganda landscape, media releases are often translated or even published in other languages, primarily English. In its ISIL Twitter census, Brookings found that three-quarters of ISIL supporters on Twitter selected Arabic as their primary language, and that out of 20,000 ISIL supporter accounts, 107 countries were represented (Berger and Morgan, 2015). Zelin's analysis of an entire week of ISIL media releases found that all 123 releases were published in Arabic; 8 of these were translated into English, and a handful were also translated into Russian, Kurdish, French, and Urdu. He also found that picture albums and videos, which can partially transcend language barriers, accounted for 88 percent of the media releases (Zelin, "Picture or It Didn't Happen").

Scholars have found that themes of ISIL propaganda can vary along with its target audience. Winter's in-depth study of ISIL propaganda distilled its messaging into six overlapping narratives: brutality, mercy, victimhood, war, belonging, and utopia. He argued that while brutality dominates the attention span of Western media, utopianism and belonging are what attract new recruits (2015). ISIL apparently hopes that brutality deters international publics from becoming active opponents, that mercy appeals to active opponents and encourages them to become potential recruits, and that belonging and utopianism lure potential recruits into becoming disseminators or active members. Zelin's 2015 study analyzed the distribution of ISIL messaging along eleven themes, the six most prevalent of which, in descending order of prominence, were military, governance, *da'wa* (call, or evangelism), *hisba* (enforcement of adherence to Islamic principles), promotion of the caliphate, and attacks by the enemy (Zelin, "Picture or It Didn't

Happen"). Many ISIL productions blend these themes into potent, multilayered messages, facilitating ISIL's attempts to achieve its strategic objectives.

Systematic and quantitative analysis of the variance between ISIL's English, Arabic, and other language productions has been lacking, though the phenomenon has occasionally been examined qualitatively. For instance, Rita Katz of the SITE Intelligence group noted differences in how the ISIL media releases in different languages reported on the San Bernardino attacks in December 2015. Several days after the attack, the various language editions of al-Bayan Radio news bulletins reported on the attack; but while the Arabic edition called the attackers, Tashfeen Malik and Syed Farooq, "supporters" of ISIL, the English edition referred to them as "soldiers" of ISIL, the Russian edition called them "mujahideen," and the French edition called them "sympathizers." The English version in particular suggested a stronger message of association and approval than the Arabic and French editions, which distanced ISIL from the attackers.

Scott Shane and Ben Hubbard have gone deeper in their qualitative study of the subject, arguing that the message of ISIL's English-language media productions is "far softer" than that in its Arabic-language media, which "flaunts violence toward its foes," especially Shi'a elements in Iraq and Syria, with videos that "linger on enemy corpses and show handcuffed prisoners casually machine-gunned." Shane and Hubbard assert that, in contrast, ISIL English-language productions emphasize jihad "as a means of personal fulfillment." In one such video, a British fighter asks, "Are you willing to sacrifice the fat job you've got, the big car, the family?" and then answers his own question: "Living in the West, I know how you feel—in the heart you feel depressed. . . . The cure for depression is jihad" (2014).

Lexical Analysis of ISIL Propaganda in Arabic and English

Given the gap in the current literature, RAND undertook a quantitative analysis of how ISIL's propaganda varies across language barriers. The English and Arabic text corpora used in the below analysis was composed of ISIL media released in both English and Arabic between

February 1 and December 9, 2015.[4] The approach that emerged as most effective for identifying distinctive keywords was comparing the English official corpus to FROWN, a baseline corpus representative of general-use English, and then separately comparing the Arabic official corpus (which had previously been translated to the English corpus) to the same baseline FROWN corpus.[5]

The analysis found that ISIL's English-language and Arabic-language propaganda overlap substantially on some themes but differ markedly on other critical themes in both substance and treatment. Both the English and Arabic propaganda place strong emphasis on ISIL's status as a territory-holding state and "caliphate," with many over-present terms like "caliphate," "lands," and the name of Abu Bakr al-Baghdadi, as well as basic Islamic religious principle and concepts, such as God, the Prophet Muhammad, and *shari'a* or religious law. However,

---

4 This method highlights similarities and differences between corpora by calculating log likelihoods for the over-presence or under-presence of keywords as well as identifying words that co-occur more often than would be predicted by chance, called collocates. All words from the baseline were included in the keyword comparisons, and a minimum frequency cut-off of three instances was employed for both keywords and collocates. All top collocates included in the foregoing analysis had pointwise mutual information (PMI) scores of over 12. All top keywords included in Figures 1 and 2 had log-likelihood scores, indicating that they were present in the target corpus more than would be predicted from their presence in the baseline corpus, and had over-presence log-likelihood scores of over 100, which is considered highly significant.

5 The corpus of English-language, official ISIL propaganda comprised 270,570 words from 44 media releases. These included full-length issues of *Dabiq* magazine, radio reports, statements, and video transcripts. ISIL media outlets represented included Hayat Media Center, Al-Itisam Media Foundation, Al-Bayan Radio, Furat Media Center, and the Raqqa, Jazeera, West Africa, and Ninawa Provincial Media Offices. The corpus of Arabic-language, official ISIL propaganda manually translated into English comprised 62,777 words from 73 media releases. These included statements, video transcripts, photo reports, infographics, martyrdom announcements, and speeches from ISIL's vaunted caliph, Abu Bakr al-Baghdadi, and official ISIL spokesperson Muhammad al-Adnani. ISIL media outlets represented included Furqan Media Foundation, Hayat Media Center, Al-Bayan Radio, Furat Media Center, Al-Naba' Publishing, and the Damascus, Homs, Aleppo, Khayr, Anbar, Sinai, Fezzan, Ninawa, Kirkuk, Salah al-Din, Baghdad, Furat, Diyala, Aden, Sanaa, Janoub, Dijlah, Aden, Sanaa, Hadramawt, Baraka, Barqa, Algeria, and West Africa Provincial Media Offices. An absence of Arabic-language publications as lengthy as the full-length *Dabiq* magazines accounts for the considerable difference in word counts between the two corpora.

marked differences between the two sets of propaganda emerge when it comes to violence and holy war, as can be seen in Figure 5.2 below.

The Arabic-language propaganda appears targeted more to the rank and file of ISIL supporters who need to be stirred into a frenzy, rather than a skeptical purveyor who needs to be won over, or a foreign government that needs to be logically rather than emotionally deterred. Arabic-language productions focus more than their English-language counterparts on bloody, warlike concepts, such as "soldiers" and "body parts," as well as "defeated" and "destroying." This different emphasis is echoed in the greater proportion of lofty, jihad-related keywords and collocates—or words that co-occur more often than would be predicted by chance—in the Arabic-language media releases, including collocates such as "mujahideen" and "martyrdom."

The Arabic-language material also includes more emotive and religiously tinged positivity, with terms like "honor," "victory," and "glad tidings," and intense, urgent language like "final outcome." The appeal to followers is also more personal, with substantially more and varied pronouns; it speaks directly to its audience with "you" and "your," ingratiates itself with its audience by referring to "we" and "us," and separates the audience from its enemies, referring to "they" and "their."

In turn, ISIL's English-language propaganda gives more prominence to coverage of "the enemy," potentially suggesting that it is targeted more toward local, regional, and foreign potential opponents that it seeks to either deter or provoke. There are more mentions of Arab and Middle Eastern governments opposed to ISIL, such as Iraq and Egypt; Eastern nations and individuals opposed to ISIL, such as Russia and Japan; and violent jihadist organizations opposed to ISIL, such as the al Qaeda-affiliated Jabhat al-Nusra.

The numbers of top keywords and collocates related to ISIL's regional "Shi'a enemies," such as Iran, the Iraqi government, and the Syrian regime are marginally higher for the Arabic-language propaganda, and references to countries and institutions of the West are roughly equal between the two sets.

Figure 5.2
Highlighted Differences: Top Keywords and Collocates by Coded Theme[6]

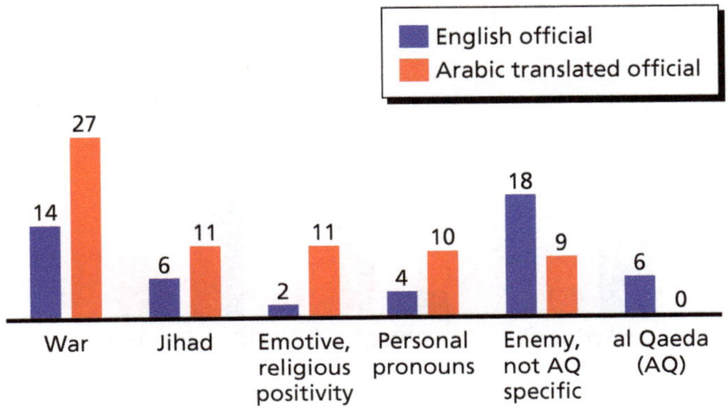

SOURCE: RAND.
NOTE: Y-axis refers to number of ISIL media communications.
RAND *RR1772-5.2*

The English-language releases represented in Figure 5.2 display twice the emphasis of the Arabic-language releases on enemies farther afield, such as "Jews," "infidels," "hypocrites," and "Crusaders," a term that ISIL sometimes uses to slander both Easterners and Westerners. This more international focus, as opposed to a focus on the specific region of Syria and Iraq, is borne out in Figure 5.3 below, which shows how the main keywords and collocates of the two propaganda sets vary in terms of locations referenced.

---

6 Specifically, this figure displays the most prevalent coded themes of the top 150 keywords as ranked by log likelihood calculations and the top 30 bigrams and top 30 trigrams as ranked by frequency for each corpus, where the coded theme was markedly more prevalent in one corpus than the other. The two target corpora were the official ISIL English-language propaganda and official ISIL Arabic-language propaganda that had been translated into English, and the target corpora were compared to FROWN, a baseline corpora that is representative of general-use English. While the ranking and identification of keywords and collocates is a statistically rigorous process, the authors make no claim of statistical significance for the hand-coded themes displayed in the figure.

Figure 5.3
Top Keywords and Collocates by Coded Location[7]

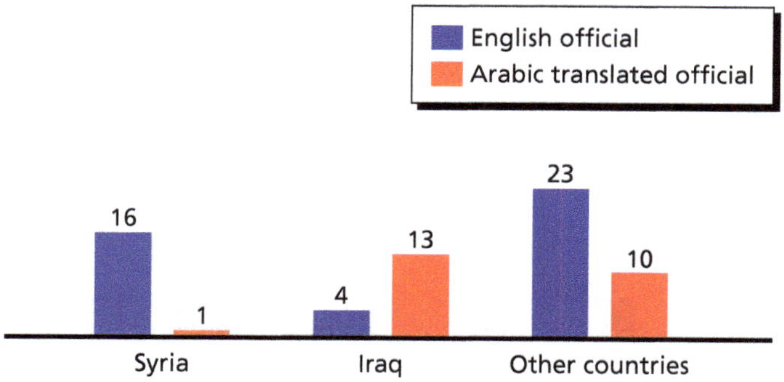

SOURCE: RAND.
RAND *RR1772-5.3*

The above figure also shows English-language media productions emphasizing Syria over Iraq. The Arabic-language media productions exhibit the opposite, focusing on Iraq more than on Syria. This imbalance may suggest that ISIL views the plight of Syria as having a broader appeal than its operations in Iraq, at least for the audiences in Western nations who are more likely to read English-language propaganda. This view may be accurate, given that the Syrian crisis was what initially sparked the unprecedented deluge of foreign fighters to the area and became a cause célèbre for Sunnis across the world, who were outraged by the international community's passivity in face of the atrocities committed by the Assad regime. Alternatively, the relative emphasis of ISIL's Arabic-language propaganda on Iraq may point toward an operational focus on Iraq. This explanation would accord with the domination of

---

[7] Specifically, this figure displays the most prevalent coded locations of the top 150 keywords as ranked by log likelihood calculations and the top 30 bigrams and top 30 trigrams as ranked by frequency for the two target corpora of official ISIL English-language propaganda and official ISIL Arabic-language propaganda that had been translated into English, compared to FROWN, a baseline corpora that is representative of general-use English. While the ranking and identification of keywords and collocates is a statistically rigorous process, the authors make no claim of statistical significance for the hand-coded themes displayed in the figure.

Iraqis among ISIL senior leadership that has been noted by some observers (Sly, 2015).

While all propaganda should be treated with caution, these findings suggest that Western governments should use a barrel rather than a teaspoon of salt when consuming ISIL's English-language propaganda. This is because, even more so than the Arabic-language propaganda, ISIL's English-language propaganda seems to be intended to manipulate the West, whether through provocation or intimidation.

Regardless of the reasons for these disparities between ISIL's Arabic- and English-language media productions, the imbalances outlined above underscore the inadvisability of focusing exclusively on one language of ISIL propaganda. Only a multilingual approach will be able to capture a holistic view of ISIL's message. For instance, an exclusive focus on ISIL's English-language propaganda, as is common in Western media, may overestimate ISIL's attention to the "far enemy."

The end result is that while ISIL's Arabic messaging, apparently targeted more to people already within the rank and file, emphasizes violence and violent jihad, ISIL's English messaging has a more international focus. This may serve the dual purpose of alternately intimidating and provoking its enemies, but also attracting more moderate supporters and potential emigrants who are perhaps more interested in the utopian aims of ISIL than in violence.

ISIL's information operations campaign weaves together almost all elements of its political warfare strategy. Its diplomatic overtures to affiliates seek to send an image of an expanding caliphate attractive to recruits. Its intelligence apparatus and violent atrocities serve to cow potential opponents into submission even while whipping its base into a frenzy. Finally, its independent sourcing of critical economic resources, as well as its sophisticated government institutions, are intended to enable the caliphate to appear strong, retain its fighters, and pacify its populace, which represents a potential recruiting pool to both ISIL and its enemies.

Finally, the study presents insights gleaned from analyzing the strengths and vulnerabilities of ISIL's multilayered information operations campaign. ISIL's strengths include its uncompromising posture as well as its embrace of structured decentralization, the latter of which might be adopted by ISIL's opponents. However, while ISIL's rebranding of itself as a caliphate has proved a tremendous boon for its recruitment, it also contains a probable message of vulnerability; selling itself as a caliphate entangles ISIL in promises that it cannot keep.

One of ISIL's key advantages in its information operations is its extremism and refusal to compromise; those qualities enable internal coherence. ISIL can easily expose inconsistencies and contradictions in American foreign policy, which teeters between realpolitik and idealism, brokers compromises between global security and the national interest, and balances the demands of international and domestic politics. The U.S. government signs deals with Iran and bombs Sunnis in Iraq, even as it supports Saudi Arabia bombing Shi'as in Yemen. This may confuse people in the Muslim world, unless they decide that the only consistency in American foreign policy that they can depend on is that America acts in its own national interests. The American message gets perceived around the world as supporting democracy, but also dictatorships. The U.S. message is complicated; it is nuanced. The United States and its allies moderate to govern. They try to strike a middle ground, and that exposes them to accusations of hypocrisy.

ISIL's message, in contrast, is simple, because ISIL does not have to care about the middle ground; in fact, it actively wants to destroy it. The seventh issue of *Dabiq* is entitled "From Hypocrisy to Apostasy: The Extinction of the Grayzone." ISIL propaganda achieves internal logical consistence by simplifying the world to a point where its worldview has no internal contradictions. Its incarnation is brand new, so it has few past actions to contradict. More than that, its extremism is absolute. It does not compromise. And that makes it impervious to nuanced argument that tries to dismantle its message from within its set of assumptions.

Even as ISIL's claim to a caliphate empowers it, it also makes it vulnerable. By declaring itself a utopia, ISIL has made promises that it cannot keep. Rising prices, food shortages, undrinkable water, and insufficient health care (Sly, 2014; Chastain, 2015) can all be leveraged as cracks in the façade of ISIL's glorious caliphate. In November 2015, a spokesman for the U.S.-led coalition against ISIL reported that desertions from ISIL are rising (Martel). ISIL defectors, especially those fighters who travelled to join ISIL but then fled, are those with the most credibility to undermine ISIL (Neumann, 2015). In order to level devastating claims of hypocrisy against ISIL, these deserters must be empowered to talk about their disillusionment with the organization.

Conclusion

ISIL's information operations are distinguished by a multilingual decentralized propaganda system, with multiple tiers of content production and dispersed propaganda dissemination via social media. This system has been made resilient by peer-to-peer sharing and deliberate lateral redundancy across multiple platforms. Additional resiliency derives from ISIL's use of disseminators outside conflict zones. Finally, official and unofficial recruiters perform targeted interventions, particularly online, in order to flatter, isolate, and radicalize recruitment targets. Unofficially affiliated hackers deface others' websites and obtain and post "kill lists" containing personal information of targets.

As for the other activities ISIL undertakes, this chapter's examination of ISIL's political warfare practices finds that its most intensive political activities that might be termed diplomatic are robust efforts to develop and support overseas affiliates and other militant jihadist organizations. However, ISIL has also engaged in tacit cooperation with some governments, primarily in order to facilitate needed imports of war materiel and fighter flows, as well as exports of oil, gas, and other sources of revenue.

ISIL demonstrates significant intelligence and counterintelligence capabilities that assist in all aspects of its operations, to include nonkinetic political warfare activity. It uses religious proselytization centers in areas of planned expansion to recruit local collectors. It monitors local tribes, resistance organizations, and even its own fighters and officials, including their Internet usage. Use of cell phones is tightly controlled, faces are concealed, and grooming standards are altered as countertargeting measures. Online communications are restricted in battle zones, and online supporters are instructed in the use of encrypted communications systems.

While ISIL has conducted maneuver warfare on a scale rarely seen by nonstate actors, its use of armed force includes systematic assassination campaigns against local leaders or organizations deemed to pose a threat, in addition to use of suicide vests and vehicle-borne improvised explosive devices (VBIEDs). The use of force is conducted with a major emphasis on amplifying its psychological effect and dissemination through media. For example, ISIL has staged hundreds of public executions and other displays of barbarism, including sexual slavery.

Although ISIL's economic activities have been carried out largely to create revenues for warfighting and to sustain its caliphate, its pursuit of these aims has also in some cases entailed political objectives. Perhaps most prominent is its effort to establish and mint its own currency as a means of demonstrating the state-like status it achieved in parts of Iraq and Syria. It has also used currency, fuel, and food as weapons by establishing, tracking, and controlling independent sources of critical resources within held territory. The extortion of domestic populations under its control may serve as a control mechanism as well as a source of revenues, as an impoverished population has less autonomy and therefore ability to escape or rebel.

One of ISIL's strengths, which the U.S. government could conceivably learn from and adapt to its own uses, is ISIL's embrace of decentralization in information operations. The U.S. State Department recently announced an end to producing its own English content and a shift to helping Arab governments shape more "localized" anti-ISIL messages to counter ISIL's message (Miller and DeYoung, 2016). While this represents a step in the direction of decentralization, it is a far cry from the sort of decentralized Twitter army that ISIL relies upon for a steady drumbeat of support. This online swarm is what enables ISIL's resilient propaganda dissemination methods, as well as the enveloping sense of belonging offered not only to alienated Muslims who emigrate to ISIL-held territory but also to those who just advocate for others to do so. ISIL taps into the desire among frustrated youth to do something real that can bring them a sense of glory and belonging, even without leaving their computer screens.

Countering ISIL effectively may require developing an equally compelling sense of community to compete with the sense of belonging that ISIL has managed to create. If anti-ISIL forces could tap into that same fervor, the effects could be significant. As ongoing research by RAND has shown,[8] the current online opposition to ISIL is fragmented by national and sectarian lines, with the most prominent ISIL opponents either shouting into isolated echo chambers or hurling insults at ISIL supporters. The idea of a U.S. government embrace of an online citizen militia may seem far-fetched, but expecting a few dozen U.S. analysts and experts to compete with thousands of highly motivated and creative ISIL supporters is equally or more unrealistic. Chapter Eight will survey the current state of U.S. strategic communications and information operations and evaluate options for improving current practices.

References—Chapter 5

Al-Masri, Abu Abdullah, "Principles in the Administration of the Islamic State" (2014), trans. Aymenn Jawad Al-Tamimi, *Guardian*, December 7, 2015. As of December 14, 2015:
http://www.theguardian.com/world/2015/dec/07/islamic-state-document-masterplan-for-power

"Ansar Bayt Al-Maqdis," Counter Terrorism Guide, National Counterterrorism Center. As of September 22, 2015:
http://www.nctc.gov/site/groups/ansar_bayt_al_maqdis.html

Barnett, David, "Ansar Jerusalem Claims Responsibility for Car Bombing in Ismailia," *Long War Journal*, October 21, 2013. As of September 22, 2015:
http://www.longwarjournal.org/archives/2013/10/ansar_jerusale_claim.php

Barrett, Richard, "Foreign Fighters in Syria," The Soufan Group, November 2014. As of December 16, 2015:
http://soufangroup.com/wp-content/uploads/2014/06/TSG-Foreign-Fighters-in-Syria.pdf

———, "Foreign Fighters: An Updated Assessment of the Flow of Foreign Fighters into Syria and Iraq," The Soufan Group, December 2015. As of January 27, 2016:
http://soufangroup.com/wp-content/uploads/2015/12/TSG_ForeignFightersUpdate3.pdf

8 Bodine-Baron et al. (2016).

Behn, Sharon, "Islamic State Oil Trade Seduces Friends, Foes Alike," *Voice of America,* December 1, 2015. As of December 15, 2015:
http://www.voanews.com/content/islamic-state-oil-trade-friends-foes/3083869.html

Berger, J. M., "Tailored Online Interventions: The Islamic State's Recruitment Strategy," Combating Terrorism Center, October 23, 2015. As of December 16, 2015:
https://www.ctc.usma.edu/posts/tailored-online-interventions-the-islamic-states-recruitment-strategy

Berger, J. M., and Jonathan Morgan, "The ISIS Twitter Census: Defining and Describing the Population of ISIS Supporters on Twitter," The Brookings Project on U.S. Relations with the Islamic World, Analysis Paper No. 20, March 2015. As of September 21, 2015:
http://www.brookings.edu/~/media/research/files/papers/2015/03/isis-twitter-census-berger-morgan/isis_twitter_census_berger_morgan.pdf

Bodine-Baron, Elizabeth, Todd Helmus, Zev Winkelman, and Madeline Magnuson, *Examining ISIS Support and Opposition Networks on Twitter*, Santa Monica, Calif.: RAND Corporation, 2016.

Carter, Joseph, Shiraz Maher, and Peter Neumann, "#Greenbirds: Measuring Importance and Influence in Syrian Foreign Fighter Networks," London: The International Centre for the Study of Radicalisation and Political Violence, King's College, 2014. As of December 16, 2015:
http://icsr.info/wp-content/uploads/2014/04/ICSR-Report-Greenbirds-Measuring-Importance-and-Infleunce-in-Syrian-Foreign-Fighter-Networks.pdf

Carter, Sara, "Islamic State Recruitment Document Seeks to Provoke 'End of the World,'" American Media Institute, *USA Today,* July 28, 2015. As of December 15, 2015:
http://www.usatoday.com/story/news/world/2015/07/28/ami-isil-document-pakistan-threatens-india/30674099/

Chastain, Mary, "Escaped Muslim Civilians: Islamic State's Utopia Is a Failure," *Breitbart*, December 3, 2015. As of January 28, 2016:
http://www.breitbart.com/national-security/2015/12/03/escaped-muslim-civilians-islamic-states-utopia-failure/

"Civilians Pay for Fuel Strikes on Islamic State," IRIN News, December 8, 2015. As of December 15, 2015:
http://www.irinnews.org/report/102278/civilians-pay-for-fuel-strikes-on-islamic-state

Comey, James, "Joint Statement with Deputy Attorney General Sally Quillian Yates before the Senate Judiciary Committee," Washington, D.C., July 8, 2015. As of January 26, 2016:
https://www.fbi.gov/news/testimony/going-dark-encryption-technology-and-the-balances-between-public-safety-and-privacy

El-Ghobashy, Tamer, "Islamic State Offshoot Entrenches in Egypt's Sinai Peninsula," *Wall Street Journal*, July 8, 2015. http://www.wsj.com/articles/islamic-state-offshoot-entrenches-in-egypts-sinai-peninsula-1436393296

Engel, Richard, "Washington Risks Falling for Syria's ISIS Strategy, Again," *NBC News*, September 14, 2014. As of January 25, 2016: http://www.nbcnews.com/storyline/isis-terror/washington-risks-falling-syrias-isis-strategy-again-n202841

Fisher, Ali, "Swarmcast: How Jihadist Networks Maintain a Persistent Online Presence," *Perspectives on Terrorism*, Vol. 9, No. 3, June 2015. As of September 21, 2015: http://www.terrorismanalysts.com/pt/index.php/pot/article/view/426/html

Fishwick, Carmen, "How a Polish Student's Website Became an Isis Propaganda Tool," *The Guardian*, August 15, 2014. As of September 20, 2015: http://www.theguardian.com/world/2014/aug/15/-sp-polish-man-website-isis-propaganda-tool

Ganor, Boaz, "Four Questions on ISIS: A 'Trend' Analysis of the Islamic State," *Perspectives on Terrorism*, Vol. 9, No. 3, June 2015.

Gidda, Mirren, "ISIS Is Facing a Cash Crunch in the Caliphate," *Newsweek*, September 23, 2015. As of December 16, 2015: http://europe.newsweek.com/isis-are-facing-cash-crunch-caliphate-333422

Hendawi, Hamza, and Qassim Abdul-Zahra, "Despite US-Led Campaign, Islamic State Rakes in Oil Earnings," Associated Press (AP), October 23, 2015. As of December 16, 2015: http://bigstory.ap.org/article/061e7a83299644868c920bed0667eb9c/despite-us-led-campaign-islamic-state-rakes-oil-earnings

Implementation of Security Council Resolution 2178 (2014) by States Affected by Foreign Terrorist Fighters, Security Council Committee Established Pursuant to Resolution 1373 (2001) Concerning Counter-Terrorism, May 2015. As of September 21, 2015: http://www.un.org/en/sc/ctc/docs/2015/N1514129_EN.pdf

Interview with Howard Shatz, "Up All Night," *BBC Radio*, November 24, 2015. As of December 15, 2015: http://www.bbc.co.uk/programmes/b06ps1w3

"ISIL Released Turks Held Hostage in Iraq," Al-Jazeera, September 21, 2014. As of January 25, 2016: http://www.aljazeera.com/news/middleeast/2014/09/turkish-hostages-seized-iraq-freed-201492035248486996.html

Kaplan, Michael, "Islamic State Going Broke? As Airstrikes Target Oil Revenue, ISIS Slashes Salaries, Imposes New Taxes," *International Business Times*, December 7, 2015. As of December 15, 2015:
http://www.ibtimes.com/islamic-state-going-broke-airstrikes-target-oil-revenue-isis-slashes-salaries-imposes-2214514

Katz, Rita, "IS Media Comments on San Bernardino Show Confusion," SITE Intelligence Group, December 5, 2015. As of December 16, 2015:
http://news.siteintelgroup.com/blog/index.php/categories/jihad/entry/407-is-media-comments-on-san-bernardino-show-confusion

Kingsley, Patrick, "Egypt Faces New Threat in Al-Qaida-Linked Group Ansar Beyt Al-Maqdis," *The Guardian*, January 31, 2015. As of September 22, 2015:
http://www.theguardian.com/world/2014/jan/31/egypt-alqaida-terrorist-threat-ansar-beyt-almaqdis

Kirkpatrick, David, "Militant Group in Egypt Vows Loyalty to ISIS," *New York Times*, November 10, 2014. As of September 22, 2015:
http://www.nytimes.com/2014/11/11/world/middleeast/egyptian-militant-group-pledges-loyalty-to-isis.html?_r=1

Klausen, Jytte, "Tweeting the Jihad: Social Media Networks of Western Foreign Fighters in Syria and Iraq," *Studies in Conflict & Terrorism*, Vol. 38, No. 1, pp. 1–22, October 17, 2014. As of December 16, 2015:
http://www.tandfonline.com/doi/full/10.1080/1057610X.2014.974948

Lando, Ben, "Mosul Insurgents Challenge Islamic State Occupiers," *Iraqi Oil Report*, July 22, 2015. As of December 16, 2015:
http://www.iraqoilreport.com/news/mosul-insurgents-challenge-islamic-state-occupiers-15337/

"Libya: More Than 100 Killed During Combats between Islamic State (ISIL) and Feryan Tribe Members in Sirte," *Wars in the World*, August 15, 2015. As of December 16, 2015:
http://www.warsintheworld.com/index.php/2015/08/15/libya-more-than-100-killed-during-combats-between-islamic-state-isil-and-feryan-tribe-members-in-sirte/

Magdy, Walid, Kareem Darwish, and Ingmar Weber, "#FailedRevolutions: Using Twitter to Study the Antecedents of ISIS Support," Qatar Computing Research Institute, Qatar Foundation, March 9, 2015.

Malik, Shiv, "The Isis Papers: Leaked Documents Show How Isis Is Building Its State," *The Guardian*, December 7, 2015. As of January 28, 2016:
http://www.theguardian.com/world/2015/dec/07/leaked-isis-document-reveals-plan-building-state-syria

———, "Support for Isis Stronger in Arabic Social Media in Europe Than in Syria," *The Guardian*, November 28, 2014. As of January 28, 2016:
http://www.theguardian.com/world/2014/nov/28/support-isis-stronger-arabic-social-media-europe-us-than-syria

Martel, Frances, "Desertion by Islamic State Fighters Is Up, Raising Fears of Attacks in the West," *Breitbart*, November 30, 2015. As of January 28, 2016: http://www.breitbart.com/national-security/2015/11/30/desertion-by-islamic-state-fighters-on-the-rise-raising-fear-of-attacks-in-the-west/

McCants, William, "How the Islamic State's Favorite Strategy Book Explains Recent Terrorist Attacks," *War on the Rocks*, November 24, 2015. As of December 16, 2015: http://warontherocks.com/2015/11/how-the-islamic-states-favorite-strategy-book-explains-recent-terrorist-attacks/

Miller, Greg, and Karen DeYoung, "Obama Administration Plans Shake-up in Propaganda War Against ISIS," *Washington Post*, January 8, 2016. As of January 27, 2016: https://www.washingtonpost.com/world/national-security/obama-administration-plans-shake-up-in-propaganda-war-against-the-islamic-state/2016/01/08/d482255c-b585-11e5-a842-0feb51d1d124_story.html

Neumann, Peter, "Victims, Perpetrators, Assets: The Narratives of Islamic State Defectors," The International Centre for the Study of Radicalisation and Political Violence, September 2015. As of January 28, 2016: http://icsr.info/wp-content/uploads/2015/09/ICSR-Report-Victims-Perpertrators-Assets-The-Narratives-of-Islamic-State-Defectors.pdf

Orange, Richard, "Jihadi John 'Forced Me to Dance Tango,' Says Former Danish Isil Hostage," October 12, 2015. As of January 25, 2016: http://www.telegraph.co.uk/news/worldnews/islamic-state/11926456/Jihadi-John-forced-me-to-dance-Tango-says-former-Danish-Isil-hostage.html

Paraszczuk, Joanna, "Mosul's Vigilante Brigades Risk It All to Take on IS," Radio Free Europe/Radio Liberty, September 22, 2015. As of December 16, 2015: http://www.rferl.org/content/islamic-state-mosul-vigilantes-take-on-militants/27263070.html

Rasmussen, Nicholas, "Countering Violent Islamist Extremism: The Urgent Threat of Foreign Fighters and Homegrown Terror," Hearing before the House Committee on Homeland Security, February 11, 2015. http://docs.house.gov/meetings/HM/HM00/20150211/102901/HHRG-114-HM00-Wstate-RasmussenN-20150211.pdf

Reuter, Christoph, "'I'm Not a Butcher': An Interview with Islamic State's Architect of Death," *Der Spiegel*, July 16, 2015. As of December 14, 2015: http://www.spiegel.de/international/world/a-conversation-with-an-isis-suicide-bomber-logistician-a-1043485.html

———, "Rational Monster: How Terror Fits into Islamic State's Plan," *Der Spiegel*, November 20, 2015. As of December 15, 2015: http://www.spiegel.de/international/world/terror-a-means-to-an-end-for-islamic-state-a-1063454.html

———, "The Terror Strategist: Secret Files Reveal the Structure of Islamic State," *Der Spiegel*, April 18, 2015. As of December 15, 2015:
http://www.spiegel.de/international/world/islamic-state-files-show-structure-of-islamist-terror-group-a-1029274.html

Robins-Early, Nick. "ISIS Debuts New Currency in Lengthy Propaganda Video," *Huffington Post*, September 1, 2015. As of October 10, 2016:
https://videos.vidible.tv/prod/2015-08/31/55e4d1aee4b070673f69bf65_640x360_LD_BASELINE_v1.mp4?MU6auQCaguxYsU81JG3kOSqui9WidrI3qK7KnV1HyjWNn-yuqCZ9IVWnLYMh2Cr7

Scarborough, Rowan, "Al Qaeda 'Rat Line' from Syria to Iraq Turns Back Against Assad," *Washington Times*, August 19, 2013. As of January 25, 2016:
http://www.washingtontimes.com/news/2013/aug/19/al-qaeda-rat-line-from-syria-to-iraq-turns-back-ag/?page=all

Shane, Scott, and Ben Hubbard, "ISIS Displaying a Deft Command of Varied Media," *New York Times*, August 30, 2014.

Shroder, Landon, "The Islamic State's Propaganda War: Advertisers and Marketers Weigh in on the World's Angriest Ad Campaign," *VICE News*, July 14, 2015. As of December 16, 2015:
https://news.vice.com/article/the-islamic-states-propaganda-war-advertisers-and-marketers-weigh-in-on-the-worlds-angriest-ad-campaign

Simpson, Cam, and Matthew Phillips, "Why U.S. Efforts to Cut Off Islamic State's Funds Have Failed," *Bloomberg Business*, November 19, 2015. As of December 15, 2015:
http://www.bloomberg.com/news/articles/2015-11-19/why-u-s-efforts-to-cut-off-islamic-state-s-funds-have-failed

Sly, Liz, "The Hidden Hand Behind the Islamic State Militants? Saddam Hussein's," *Washington Post*, April 4, 2015. As of January 27, 2015:
https://www.washingtonpost.com/world/middle_east/the-hidden-hand-behind-the-islamic-state-militants-saddam-husseins/2015/04/04/aa97676c-cc32-11e4-8730-4f473416e759_story.html

———, "The Islamic State Is Failing at Being a State," *Washington Post*, December 25, 2014. As of January 26, 2015:
https://www.washingtonpost.com/world/middle_east/the-islamic-state-is-failing-at-being-a-state/2014/12/24/bfbf8962-8092-11e4-b936-f3afab0155a7_story.html

Trumbull, Mark, "Countering Islamic State: How Much Can a Squeeze on Oil Revenue Do?" *Christian Science Monitor*, November 23, 2015. As of January 18, 2017:
http://www.csmonitor.com/USA/Foreign-Policy/2015/1123/Countering-Islamic-State-How-much-can-a-squeeze-on-oil-revenue-do-video

Winter, Charlie, "The Virtual 'Caliphate': Understanding Islamic State's Propaganda Strategy," Quilliam Foundation, July 2015. As of September 21, 2015: http://www.quilliamfoundation.org/wp/wp-content/uploads/publications/free/the-virtual-caliphate-understanding-islamic-states-propaganda-strategy.pdf

Wolf, Mat, "The Tribe That Won't Stop Killing ISIS," *The Daily Beast*, December 10, 2015. As of December 15, 2015: http://www.thedailybeast.com/articles/2015/10/12/the-tribe-that-won-t-stop-killing-isis.html

Zelin, Aaron, "New Statement from the Islamic State: 'To the Defiant Tribes of the Sinai—Wilāyat Saynā','" *Jihadology*, September 19, 2015. As of September 21, 2015: https://azelin.files.wordpress.com/2015/09/the-islamic-state-22to-the-defiant-tribes-of-the-sinai-wilc481yat-saync48122.pdf

———, "New Video Message from the Islamic State: 'Victory from God and an Imminent Conquest—Wilāyat al-Khayr,'" *Jihadology*, May 11, 2015. As of September 21, 2015: http://jihadology.net/2015/05/11/new-video-message-from-the-islamic-state-victory-from-god-and-an-imminent-conquest-wilayat-al-khayr/

———, "Picture or It Didn't Happen: A Snapshot of the Islamic State's Official Media Output," *Washington Institute, Perspectives on Terrorism*, Vol. 9, No. 4, August 2015. As of September 22, 2015: https://www.washingtoninstitute.org/uploads/Documents/opeds/Zelin20150807-Perspectives.pdf

Zelin, Aaron, Rena and Sami David Fellow, "The Decline in Islamic State Media Output," ICSR, King's College London, December 4, 2015. As of December 16, 2015: http://icsr.info/2015/12/icsr-insight-decline-islamic-state-media-output/

SITE Intelligence Group

"Blog Post Suggests Ways for IS Supporters to Boost Social Media Effectiveness," SITE Intelligence Group, July 17, 2015.

"'Cyber Caliphate' Forward Personal Information of Purported US Military Personnel," SITE Intelligence Group, November 27, 2015.

"Document Alleges IS' Attempts to "'Buy'" Ansar al-Furqan's Allegiance," SITE Intelligence Group, November 9, 2015.

"IS Shifts Propaganda Archive to the Dark Web," SITE Intelligence Group, November 15, 2015.

"Islamic Cyber Army Continues Attempted Hacking Campaign Against Iran," SITE Intelligence Group, November 10, 2015.

"Islamic Cyber Army Posts 270k Purported Twitter Account Logins for 'WorldUnderHacks' Campaign," SITE Intelligence Group, November 23, 2015.

"Jihadists Circulate Guide to Recover from Twitter Suspensions," SITE Intelligence Group, March 30, 2015.

"Pro-IS Hackers Deface Iranian Websites, Threaten to Release Personal Info," SITE Intelligence Group, November 7, 2015.

"Pro-IS Hackers Forward Public Info of French Government Personnel as Part of #WorldUnderHacks," SITE Intelligence Group, November 24, 2015.

"Pro-IS Hacking Group Claims Attack on Broadcasting Station," SITE Intelligence Group, May 13, 2015. As of January 26, 2015: https://news.siteintelgroup.com/Jihadist-News/pro-is-hacking-group-claims-attack-on-broadcasting-station.html

Cited Items from the Primary Source Archive

Archive of Islamic State Administrative Documents, assembled and translated by Aymenn Jawad Al-Tamimi, January 27, 2015. Last Updated November 10, 2015. As of December 15, 2015:
http://www.aymennjawad.org/2015/01/archive-of-islamic-state-administrative-documents

———, Specimen 4H: Free Entry to Waritheen Hotel and Amusement Park in Mosul, May 2015. Originally sourced from:
http://justpaste.it/waritheenhotelinvite

———, Specimen 4I: Statement to the Ninawa Province Tribes, March 23, 2015. Originally sourced from:
http://justpaste.it/ninawatribesl

———, Specimen 4R: Gas Distributions in Deir az-Zor Province, Al-Khayr Province, Diwan al-Rikaz, May 24, 2015. Original source:
https://pbs.twimg.com/media/CGB1a0iWMAIcBUt.jpg

———, Specimen 9A: Suspension of Internet services, Albukamal, Intelligence Services Official, Wilayat al-Furat, September 2015.

———, Specimen 9C: Prohibition on Use of New Syrian Banknotes, Diwan al-Hisba, Wilayat Halab, September 27, 2015.

———, "Specimen 5E: General Notification, Aleppo Province," General Administration, Wilayat Halab, February or March 2015. Original source:
http://justpaste.it/aleppogeneralnote

———, Specimen 8Z: Cutting Off Mobile Phone Networks, Media Office, Ninawa Province, November 28, 2014.

———, Specimen 6O: Recording IDs of Internet Café Users, Public Security Department, Raqqa Province, July 22, 2015. Original source: https://pbs.twimg.com/media/CKxVvspW8AA0Ygg.jpg

———, Specimen 2P: Shari'a Committee for Observation of New Moons (Setting Up New Calendar), Sharia Committee for the Observation of New Moons, December 18, 2014.

———, Specimen 5U: Another Oil and Gas Sales Receipt, Kirkuk Province, Oil and Gas Division, Diwan al-Rikaz, January 2015. Original source: http://justpaste.it/kirkukoilsales5

———, Specimen 8C: Form for Dealing with the Affairs of the Tribes (Albukamal area), Committee to Track the Affairs of the Tribes.

"The Blocklist Tutorial," Ansaru Khilafa, March 26, 2015. As of December 1, 2015: https://ansarukhilafah.wordpress.com/2015/03/26/the-blocklist-tutorial/

Dabiq: The Failed Crusade, Al-Hayat Media Center, No. 4, October 11, 2014.

Dabiq: Remaining and Expanding, Al-Hayat Media Center, No. 5, November 21, 2014.

Dabiq: Shari'ah Alone Will Rule Africa, Al-Hayat Media Center, No. 8, March 30, 2015.

Dabiq: They Plot and Allah Plots, Al-Hayat Media Center, No. 9, May 21, 2015.

"How to Use GPG4USB for Secure Secure Encrypted," Ansaru Al-Khilafa, April 6, 2015. As of December 1, 2015: https://ansarukhilafah.wordpress.com/2015/04/06/how-to-use-gpg4usb-for-secure-encrypted-communications/

"ISIS Abu Bakr al-Baghdadi First First Friday Sermon as So-Called 'Caliph,'" *Al-Arabiya*, July 5, 2014. As of January 25, 2016: http://english.alarabiya.net/en/webtv/reports/2014/07/07/ISIS-Abu-Bakr-al-Baghdidi-first-Friday-sermon-as-so-called-Caliph-.html

"Registering on Twitter Without Phone Verification, Part 1," Ansaru Khilafa, March 26, 2015. As of December 1, 2015: https://ansarukhilafah.wordpress.com/2015/03/26/registering-on-twitter-without-phone-verification-part-1/

"Registering on Twitter Without Phone Verification, Part 2," Ansaru Khilafa, March 26, 2015. As of December 1, 2015: https://ansarukhilafah.wordpress.com/2015/03/26/registering-on-twitter-without-phone-verification-part-2/

CHAPTER SIX

Attributes of Modern Political Warfare

The preceding case studies have shown that each actor—be it the United States, Russia, Iran, or ISIL—employs its own unique form of warfare below the threshold of conventional war. This is by no means an exhaustive list. At least two states not covered in this study deserve special mention—Great Britain and China—since the former was at the forefront of political warfare during the nineteenth and first half of the twentieth century; and the latter, as noted in this chapter's distillation of current practices, is actively engaged in political warfare today. While each of these actors has its idiosyncrasies, some broad commonalities exist in the practice of political warfare today. This chapter draws on the lessons from the preceding ones to discuss eight attributes that help define the modern practice of political warfare.

1. Nonstate actors can conduct political warfare with unprecedented reach.

The idea that nonstate actors can engage in political warfare is not particularly new, nor is the idea that they can have global impact. In the late nineteenth and early twentieth century, Anarchists, Marxists, and other groups managed to organize, communicate, and operate across borders (Hoffman 2006, pp. 7–11). Yet nonstate and pseudo-state actors today have more power to influence the international system than ever before. As Richard Haass claimed almost a decade ago, "One of the cardinal features of the contemporary international system is that nation-states have lost their monopoly on power and in some domains their preeminence as well" (Haass, 2008, p. 45). Thanks to globalization, the world is increasingly interconnected, giving new power to nonstate actors. As John Arquilla and David Ronfeldt argued in their classic study *Networks and Netwars*, "The rise of networks means that power is migrating to nonstate actors, because they are able to organize into sprawling multiorganizational networks (especially 'all-channel' networks, in which every node is connected to every other node) more readily than can traditional, hierarchical, state actors" (Arquilla and Ronfeldt, 2001). And this, in turn, means that nonstate actors have new abilities to wage political warfare.

For nonstate actors, political warfare offers a chance to achieve their objectives, even if they lack the capability of waging conventional or nuclear war, and the advent of modern technology and globalization has lowered the barriers for nonstate actors wishing to engage in this form of conflict. The Internet allows for nonstate actors to easily spread propaganda messages across borders. With about 875 million guns already in circulation (only about a quarter of which are controlled by military, police, or other government agencies) and an additional 7.5 to 8 million small arms—portable, individual firearms (e.g., rifles, pistols, and light machine guns)—produced globally each year, nonstate actors have increased access to weapons and can now more readily engage in low-end military conflict (Al Jazeera, 2013; Peters, 2009). Modern communications and encryption technology further ease the means to conduct "covert diplomacy"—should a nonstate actor choose to do so. Global financial flows enable nonstate actors to transfer money to various proxies all over the world and hide these transactions.[1] In short, nonstate actors have the ability to wage political warfare across all four dimensions of power—diplomatic, informational, military, and economic.

Nonstate actors today capitalize on these newfound capabilities. Corporations are using at least some of political warfare's tools (short of military) to serve their economic interests, while ISIL actively engages in this form of warfare to promote its jihadist ideology. And they are not alone. Before ISIL, al Qaeda created a multinational Salafist

---

[1] For an overview, see Department of Treasury, 2015.

jihadist organization that leveraged Arab state weakness, anti-U.S. sentiment, and global networks to recruit, train, and facilitate its operations, financed by a multinational donor base and web of madrasas. Its appeal spawned numerous global affiliates, including al Qaeda in the Arabian Peninsula, al Qaeda of the Islamic Maghreb, Al Shabaab, Abu Sayyaf, Jemaah Islamiyah, and others (Bajoria and Bruno, 2012). Lebanese Hezbollah (LH) has evolved from a proxy force trained, armed, and directed by Iran to a social-political-military entity to a quasi-state force with a role in Lebanon's government and with hybrid military capabilities that have been projected beyond its borders into Israel, Syria, and even South America (Masters and Laub, 2014). Colombia's FARC and Sri Lanka's LTTE also managed to expand from local actors engaged in small-scale conflicts to recruit for and finance their operations from globalized networks (Hoffman, 2006, pp. 204–6; "Revolutionary Armed Forces of Colombia," 2015). And so, today, even relatively small nonstate actors have the ability to conduct political warfare on the global level.

2. Political warfare employs all the elements of power.

Political warfare is an inherently complex task: Its successful application often employs a combination of all the DIME elements of national power. Indeed, one of the striking findings of the preceding cases studied is just how rarely any actor—state or nonstate, democracy or authoritarian regime—relies on only one element of national power to conduct political warfare.

The United States used a mixture of diplomatic ventures, propaganda, armed proxies, foreign aid, and sanctions as it waged the Cold War, just as Russia and Iran use many of these same combinations—albeit with somewhat different flavors—in Ukraine and Syria, respectively, today. ISIL, too, combines propaganda and economic subversion (although more defensively to acquire weapons and resources than as a form of leverage against others abroad), along with terrorist activities in its quest to expand its influence.

At the same time, as political warfare often requires coordination across all facets of national power, states have historically tended to create specialized agencies to handle niche aspects of this form of warfare. Covert military operations have become the specialty of the Iranian Revolutionary Guard Corps-Qods Force, the Russian Spetsnaz, and the U.S. Central Intelligence Agencies' paramilitary wing. Some states set up specialized government agencies to handle the nonlethal dimensions of political warfare as well. Leaving aside its more indirect proxies, the Russian government controls the Rossiya Segodna as its official international informational agency.[2] Iran likewise has the Islamic Republic of Iran Broadcasting (IRIB), which controls all Iranian broadcasting (Pahlavi, 2012). Up until a few decades ago, the USIA before dissolving it in 1999. Earlier, during World War II, Britain had created the Political Warfare Executive, an organization subordinate to the Ministry of Foreign Affairs and the Ministry of Information, to oversee propaganda—from leaflets to radio broadcasts (Overy, 2015, p. 382; PBS). Ultimately, this creates a tension in the conduct of political warfare: It often requires a broad-based approach, but the tools for conducting this form of warfare tend to be highly specialized.

Perhaps, given this tension, many countries have needed to adapt their government structures to coordinate this form of warfare. Russia recently founded the National Center for the Management of Defense of the Russian Federation under the General Staff to help coordinate military efforts along with other agencies—such as the Ministry of Internal Affairs and Ministry of Emergency Situations (McDermott, 2014). For a time, China centralized its conduct of political warfare within the military. Some analysts even labeled the People's Liberation Army's General Political Department Liaison Department (GPD/LD) as "PLA's principal political warfare command" (Stokes and Hsiao, 2013, p. 3). Even ISIL integrates its conduct of political warfare by embedding an economic targeting officer and an intelligence service officer (to assist with information efforts) down to the district level (see Figure 5.1). Whether or not an overarching separate coordinat-

---

[2] Rossiya Segodna was established on December 9, 2013, by President Putin's decree on "Measures to increase effectiveness of state media." See RIA Novosti, n.d.

ing body is indeed necessary for the successful conduct of political warfare is beyond the scope of this study, but clearly one of the central challenges of those seeking to conduct political warfare is keeping the approach synchronized and the variety of agencies pointed in the same direction.

The application of political warfare methods can occur in numerous situations. They may include a low-cost method of competing with and cowing adversaries in ways that achieve all of the intended objectives. They may be used, as the historical survey illustrated, in a "hybrid" mode combined with more conventional use of military power. They may be a precursor to traditional warfare. In an era of growing near-peer competition and empowered nonstate actors, the use of political warfare to gain advantage may increase, because the dominance of any single actor has diminished. As defense expert and former Undersecretary of Defense Fred Ikle noted at the end of the Cold War, "Nor can we forget that POLWAR and PSYOP came back into prominence because of the relative military strengths of the contending powers. Without effective deterrence, POLWAR cannot protect us from unleashed might, whether conventional or nuclear" (Ikle, 1989, p. 10). The same may be true today. America's adversaries turn to political warfare because they find it a relatively effective and low-cost form of competition that does not incur the risks of confronting the United States head on. The United States, in turn, uses political warfare because it offers additional options to policymakers.

3. Political warfare relies heavily on proxy forces and means.

Political warfare often is conducted in the shadows. Across the cases discussed here, many of the major organizations charged with conducting political warfare—Russia's *spetsnaz*-GRU and FSB, the United States' CIA, and Iran's IRGC-QF—remain shrouded in secrecy. Often, the visible faces of political warfare are either the proxies—friendly political parties, nongovernmental organizations, loyal ethnic groups, and the like, who appear as independent but are actually backed, if not controlled by, these states from behind the scenes—or else quasi-government agencies, like state-aligned corporations. This is not a new development, either. In an earlier era, the government-aligned British East India Company helped conquer and rule India for Britain. To a lesser extent, in the late nineteenth and early twentieth centuries, the private United States Fruit Company helped expand American control over the Caribbean (Omang, 1985, p. 5). While the British East India Company and United States Fruit Company may be gone, the practice of state-aligned corporations becoming tools of the state to exert leverage abroad remains alive and well today.

Russia maintains a spectrum of control over organizations—from direct control through loose affiliation—in order to wield power (see Figure 3.1). These cutouts perform a range of political warfare activities, from information operations to economic coercion. For example, Russia owns a majority stake in energy giant Gazprom, one of the major exporters to the countries of the former Soviet Union, and Russia has leveraged this energy dependence to attempt to wield influence abroad (Abdelal, 2013).

China also extensively uses cutouts—seemingly private business and cultural entities that are in fact controlled by the Chinese government and the PLA. According to the Asia-focused Project 2049 think tank, the number of civilian proxies for the GPD/LD has been growing and now includes the China Association for Promotion of Chinese Culture, China Association for Friendly International Contacts, China-U.S. Exchange Foundation, The Centre for Peace and Development Studies, External Propaganda Bureau, and China Energy Fund Committee (Raska, 2015; Parameswaran, 2015). According to China analyst J. Michael Cole, "The convoluted nature of the network—where global energy firms overlap with think tanks, foundations, the GPD/LD and intelligence operatives—makes it extraordinarily difficult for intelligence agencies (and the less security-aware targets) to distinguish between political warfare and more 'legitimate' activities" (Cole, 2015).

In Western democracies today, state-aligned corporations are still very much part of the political warfare arsenal. State-run industries—particularly in the energy and defense sectors—exist in many Western democracies. Even the United States, with its strong history of free enterprise, still uses private and semi-private corporations to include various defense contractors to gather information and provide military assistance—all to serve its political objectives (see Singer, 2007; Shorrock, 2009; Cohen, 2010).

The use of proxies offers a host of potential strategic benefits. A propaganda message may seem more credible if it originates from a seemingly disinterested—and hence presumably objective—third party. Proxy military forces may offer the potential of added strategic surprise, better knowledge of local dynamics, and the possibility for the backer to walk away more easily, if something goes wrong (see Salehyan, 2010). Back-channel diplomacy may allow for a degree of candor with nefarious actors, otherwise impossible if the negotiations were to occur in the public eye.[3] Seemingly private corporations may be regarded with less suspicion and enjoy more access than official state entities.

In twenty-first-century political warfare, the plausible deniability that comes from using unattributed means offers another strategic advantage: entrapping the target in a legal quagmire. Retired Air Force major general and staff judge advocate turned Duke law professor Charles Dunlap coined the term "lawfare." In twenty-first-century warfare, Dunlap argues, "law has evolved to become a decisive element—and sometimes *the* decisive element—of contemporary conflicts" (Dunlap, 2009, p. 34). He notes that casting legitimate actions as illegal can undermine all-important public support (Dunlap, 2009, p. 35). Arguably, while lawfare can be applied to any actors, Western democracies and others who respect a rules-based approach to international order are perhaps particularly vulnerable and this, in turn, is one of the reasons why these methods are so appealing in political warfare.

The use of proxies or other unattributed means by potential adversaries to slow—if not undermine—the West's response. For example, one of the concerns about the use of political warfare by Russia against the Baltics stems from differing Allies' perceptions about what would clear the policy bar for an Article 5 response—the collective-defense clause of NATO. As the *Economist* recently pointed out, attribution problems may mean, "What might count locally as an intolerable

---

3 See a discussion of clandestine diplomacy, see Scott, 2004.

assault on the Baltic states' sovereignty may not be seen in Brussels as an 'armed attack' for Article 5 purposes" ("How NATO's Article 5 Works," 2015). Other analysts identify attribution problems as one of the core challenges behind a state's ability to respond to cyber attacks.[4] Even if states eventually solve the attribution problem in political warfare, it often takes time for the intelligence to be gathered, the legal arguments developed, and the decision to respond made—all of which make unattributed forces an appealing tool in the conduct of political warfare.

4. The information arena is an increasingly contested battleground, where perceptions of success can be determinative.

Thankfully, the twenty-first-century democracies have an additional tool in the battle against modern political warfare: The ubiquity of independent media aided by modern technology that can at least shed light on some secretive political warfare tactics. While print media outlets have declined in size and staff over the last decade, there has been a dramatic growth in the digital news industry (Jurkowitz, 2014). Among the larger digital news providers, the Yahoo-ABC digital network attracts some 127,995,000 viewers; CNN.com, 101,540,000; and NBC News Digital, 101,145,000 (Olmstead and Shearer, 2015). Moreover, the numbers of providers are growing. A Pew Study found that of its sample of 438 small news outlets designed as digital news providers, 30 percent started since 2010, and 85 percent started since 2005, and most of them filled gaps in local news and investigative journalism left by larger producers (Jurkowitz, 2014).

Arguably as important as the growth of digital outlets is the growth of informal media. More than a decade ago, Lucas Graves argued in *Wired Magazine* that, in 2005, "everyone's a reporter." He noted, "These days, 'our man on the scene' is often a swarm of amazingly prolific nonprofessionals posting up-to-the-minute stories and

---

[4] For a discussion of this, see Tsagourias, 2012.

pictures of breaking news from their laptops" (Graves, 2005). If anything, Graves's observation is even truer more than a decade later. Today, smartphones come with ever-better digital cameras and Internet connections, and thanks to social media, potential stories can spread around the world near instantly.

This growth in the number of reporters and outlets works in favor of those wanting to expose covert political warfare. The media played a role in helping to expose American attempts at covert military support during the Iran-Contra scandal and at aiding the Afghanistan mujahedeen during the 1980s. More recently, the media helped unmask Russia's "little green men" in Ukraine, identifying them as GRU operatives (Rosenberg, 2014; Shevchenko, 2014). It should come as little surprise, then, that shortly after assuming control of Crimea, Russia decided to crack down on open media (Freedom House, 2014). Perhaps, though, the more important contribution of open media in unmasking political warfare's shroud of secrecy is the more mundane type of reporting—revealing political corruption, payoffs, and foreign government influence.

At the same time as independent media are countering—or at least exposing—political warfare, actors today are investing heavily in information warfare as part of their political warfare strategies. Iran's IRIB is in 45 countries with five foreign-language channels (Pahlavi, 2012, pp. 21–33). Within the last two decades alone, it launched the 24-hour Arabic TV channel Al-Alam (in 2003), the English-language channel PressTV (in 2007), and the Spanish language HispanTV (in 2011) (Pahlavi, 2012, pp. 21–23; Black, 2010). Russia, too, invests heavily in propaganda. According to some estimates, Russia funds the Rossiya Segodya news and broadcasting company to the tune of $170 million annually (Tétrault-Farber, 2014) while it maintains the affiliated RT network—including RT International (flagship channel in English), RT America (English), Rusiya Al-Yaum (Arabic), Actualidad RT (Spanish), and the documentary channel RTDoc—with a 2015 budget (according to RT) of $225 million (Shuster, 2015). China similarly has ramped up its propaganda effort. Based out of Nanjing Military Region's 311 Base (sometimes referred to as the Public Opinion, Psychological Operations, and Legal Warfare Base) in Fuzhou City,

Fujian Province, the PLA has broadcasted the "Voice of the Taiwan Strait" radio since the 1950s but recently expanded to other mediums as well, including social media, publishing, and business networks (Raska, 2015).

Western democracies also invest in information warfare, although perhaps to a lesser extent. The United States still maintains the Voice of America (VOA) and Radio Free Europe/Radio Liberty (RFERL). Similarly, the United Kingdom still has the British Broadcasting Corporation (BBC), whose express purpose is to "represent the UK, its nations, regions and communities" and to "bring the UK to the world and the world to the UK" (BBC, n.d.). The BBC, VOA, and RFE/RL, however, assert their editorial independence from their parent governments. Still, thanks to the wars in Afghanistan and Iraq, on a tactical level at least, the United States has renewed its focus on information operations.

Even nonstate actors are investing heavily in information operations. ISIL releases messages through multiple media organizations; issues an English-language magazine *Dabiq*, French-language magazine *Dar al-Islam*, and Turkish-language magazine *Constantinople*; and maintains the A'maq News Agency and Al-Bayan Radio Station. The Shi'ite terrorist organization Hezbollah has a long-standing interest in media. It established its first newspaper in 1984 (*Al-Ahd*), its first radio station (*Al-Nour*) in 1988, and its first television channel Al-Manar TV in 1991 (Hashem, 2015).

Today's increasingly heated battle for control of the information domain fits neatly with the historical conception of political warfare. Some of the initial scholarship on political warfare conflated the term with psychological warfare. The focus on information warfare also is in keeping with the idea that political warfare—like many other forms of irregular warfare—is principally a battle over hearts and minds. There is no shortage of quotes—from classical British counterinsurgency theorist Frank Kitson to al Qaeda leader Ayman al-Zawahiri—extolling the importance of the psychological dimension and identifying psychological warfare as one of the most important elements, if not the most important element, in irregular warfare campaigns.[5] Judging simply on the inputs and the resources—both time and money—that they have sunk into information warfare, most of the actors covered here seem to agree.

Information warfare is an increasingly important battleground in modern political warfare in which the media can function as a double-edged sword. While independent media can help uncover political warfare, many actors are investing heavily in trying to control media flows. At the moment, it is impossible to say with any certainty—at least in the unclassified setting—the precise number of political warfare activities that remain covert and perhaps, more importantly, what percentage of covert political warfare activities are exposed by media coverage. As a result, it is impossible to say conclusively who is winning—the actors who want to conduct covert political warfare or those who want to expose it.

5. Information operations create effects in various ways by amplifying, obfuscating, and, at times, persuading. Factual evidence supplied in a timely manner is the best antidote to disinformation.

The ubiquity and instantaneous nature of communications provides all actors with opportunities to spread their messages worldwide and attempt to gain influence or leverage through this means. A variety of information and psychological effects have been achieved by ISIL, on a wide array of audiences, for example: Its brand has eclipsed that of al Qaeda in name recognition and sheer quantity of messaging; it has raised the largest multinational terrorist army in recent times, largely through its Internet recruitment and informational campaign; and it has successfully evaded attempts to squelch its message and track

5 "The main characteristic which distinguishes campaigns of insurgency from other forms of war is that they are primarily concerned with the struggle for men's minds" (Kitson, 1973); "I say to you: that we are in a battle, and that more than half of this battle is taking place in the battlefield of the media. And that we are in a media battle in a race for the hearts and minds of our Umma" (Ayman al-Zawahiri to Abu Musab al-Zarqawi, July 9, 2005).

its digital footprint. Information operations can be seen as the accelerator or force multiplier of all other elements of power, magnifying acts and actors through sophisticated Potemkin subterfuge. If these measures fail to persuade or produce converts, they can certainly confuse, distort, and misinform in ways still useful to the actor's ultimate objectives.

It is worth noting that even with today's increased investment in information warfare, the cost of wielding this element of power still pales in comparison to other forms of coercion. Conventional military might come with a far higher price tag. The Russian defense budget is estimated at $50 billion (over 100 times what it spends on RT and Rossiya Segodya combined), and its airstrikes in Syria alone came with an estimated price tag of $4 million a day (Hobson, 2015). Economic sanctions can prove equally costly. The European Union's sanctions against Russia are believed to have cost the EU some €40 billion in 2014 and €50 billion in 2015, a negative drag on its economy of 0.3 percent and 0.4 percent of its GDP, respectively (Szczepański, 2015, p. 4). Viewed in this light, information warfare is comparatively cheap.

Moreover, information warfare comes with fewer risks and a wider set of benefits. Unlike covert military action, information warfare seems to pose less of a risk of a conflict escalating into a full-fledged war. Unlike economic sanctions, information warfare does not carry the risk of hurting the sender's own economy and consequently does not come with the same domestic political risks. And unlike covert diplomacy, information warfare offers an opportunity to affect a wider audience and thus have a larger strategic impact. For both financial and strategic reasons, it is easy to understand the lure of information warfare today.

Nonetheless, when examined closely, these case studies present a mixed finding about information warfare's effectiveness in the modern age. In the case of Russian information operations, the analysis suggests that much of Russian propaganda—including relatively expensive endeavors like RT—may neither be as well-watched nor well-believed as some might presume. Similarly, American propaganda efforts during the Cold War may have played an important morale-building function, but their ultimate effect on the conflict remains unclear. By contrast,

Iran has used social media and propaganda to recruit Shi'ite forces aggressively in both Iraq and Syria (Smyth, 2015; "Iran Quds Chief Visited Russia Despite U.N. Travel Ban," 2015). ISIL has leveraged information operations to recruit and radicalize. Russia also has shown a capacity for disinformation—at the very least, they can confuse audiences about their actions and intentions. Russia can mislead international audiences (at least for a time) about the presence of its soldiers in Ukraine (see Giles, 2016, p. 37).

This mixed finding hints at four major trends shaping modern information warfare's effectiveness. First, information warfare starts at home and can prove quite effective at its domestic function: ensuring regime stability. In some cases, the state can still control the media, advances in modern technology notwithstanding. In Russia, the state controls much of the media, particularly in the television and radio spheres (Krasnoboka, n.d.). Other actors, like ISIL, may have less technical capacity to control the number of radio or television stations, but can compensate for this deficiency with brutality, allowing their propaganda to convey a simple, vivid—if violent—message about the domestic consequences of defying the Islamic state.

Second, in contrast, the democratization of information generally undercuts information warfare's ability to persuade external audiences. The spread of the Internet and the rise of multiple small media outlets allow people to tap into more sources of information than ever before. From an information warfare standpoint, this increases the challenge of selling positive messages, especially to skeptical audiences, since whatever the propaganda message is, it has to compete with a variety of alternative and oftentimes more trustworthy sources. For example, RT finds itself challenged to compete with other news outlets in the West, such as BBC or CNN, which routinely draw higher ratings on both television and on social media.

Third, however, modern technology makes it easier to reach *sympathetic* external audiences and to amplify the impact of the political warfare's diplomatic, military, and economic dimensions. Increased access to information cuts both ways: Just as it is easier for the average person to draw on information from benign sources, people can also more readily access propaganda material, too—especially if they are searching for it. Thus, the democratization of information sources acts only as a check on propaganda *if* the individuals in question are curious enough to seek out alternative viewpoints.

Fourth, and perhaps most troubling, modern technology makes disinformation easier. As Keir Giles argues in a recent analysis of Russia's disinformation attempts, "By comparison with the pre-Internet era, the effective seeding of disinformation is now vastly simpler. Noisy and unsubtle exploits such as hacking the Twitter feed of a major news agency to plant disinformation have taken place, but even these are unnecessary when stories can be introduced into the media by other, seemingly natural and legitimate, means" (Giles, 2016, p. 26). According to Giles, while Russia's disinformation campaign may seem "clumsy, counterproductive, obvious and easily debunked," the campaign may be succeeding at two other goals: controlling the flow of information at home, while obfuscating the truth abroad (Giles, 2016, p. 32). Through a variety of means, from trolls to bots to state-run media networks, and aided by Western media's own interest in "balanced" reporting, Russia hopes to "influenc[e] mass consciousness, and creating an environment in which it is hard to distinguish quality information, in order to undermine the objectivity of Western media reporting and hence influence the information available to policy-makers" (Giles, 2016, p. 33). Ultimately, Russia may not succeed at convincing the majority of Western audiences that its viewpoint is right, but it may succeed in making sizeable swaths of the population doubt the truth. Arguably, the same applies to the other cases as well.

It is too simplistic to say that information warfare "does" or "does not" work in modern political warfare. Watching RTV and PressTV or reading *Dabiq* may not turn most audience members into supporters of Russia, Iran, or ISIL. Still, information warfare can certainly help preach to the converted and perhaps sway those already somewhat sympathetic to the cause. More broadly, disinformation campaigns can help "muddy the waters" and confuse the truth. For many actors, these latter functions may slow or paralyze Western decisionmaking and thus provide a sufficient victory.

For democracies, the best antidote to disinformation campaigns is compelling factual evidence supplied in a timely manner. In his

1989 essay on political warfare (POLWAR) and psychological operations (PSYOPS) in the "modern context," Fred Ikle writes, "truth is democracy's best POLWAR and PSYOP weapon" because "the goals of democracy can only be accomplished with methods compatible with democracy" (Ikle, 1989, p. 7). Ikle and others argue that democracies have faced unique challenges when conducting political warfare partly because a democratic government cannot wholly control the flow of information. As Stanfield Turner noted, investigative journalism by the 1980s had already curtailed the U.S. government's ability to employ covert propaganda abroad and to secretly fund guerilla groups (Turner, 1985, p. 175). Even bracketing the freedom of press issues, the U.S. government has routinely struggled to speak with a unified voice. Senators and congressmen speak their minds, as do soldiers, civil servants, and even political appointees. The essence of a democracy is not just freedom of press but also freedom of speech. As a result, the U.S. government's ability to coherently "spin" situations has been limited (for an example of this argument, see Ikle, 1989, p. 6; and Lord, 1989, p. 25). If anything, the advent of the Internet, social media, and the general ease of sharing information has only increased the challenges of keeping secrets and sending coherent, cohesive messages. As such, rather than risk getting caught in a lie, it has generally been more prudent for the United States and other democracies to avoid deceptive spin altogether and to rely on the truth as their best weapon.

6. Detecting early-stage political warfare requires a heavy investment of intelligence resources.

Political warfare places a high demand on intelligence, for two reasons. First, knowing what political messages will resonate is challenging enough in domestic politics; knowing what messages will resonate with a foreign audience proves even harder. Second, political warfare often must also confront principal-agent problems—that is, where the local political and military groups may have different objectives than the United States. As a result, even when political warfare is tactically successful—as it was in backing the Christian Democrats in Italy and the mujahedeen in Afghanistan—the final outcomes may not be what the United States desired. Even more problematically, the United States can find itself without an agent that aligns with its intent or that it can control. As Codevilla notes, "The search for the ever-elusive 'moderate faction' seems to have become a substitute for knowledge about, and leverage over, foreign situations" (Codevilla, 1989, p. 91). While Codevilla was describing the American attempts to find a suitable anti-apartheid group, the same argument can be made about more recent American attempts at political warfare—such as the failed attempt to build a moderate Syrian army.

Detecting and then countering political warfare requires detailed, accurate intelligence. As Hans Speier wrote in one of the early works on political warfare, "If a response by action is the objective of political warfare and the test of its success, propaganda must be directed with a clear understanding of what kind of action is both practical and politically desirable" (Speier, 1948, p. 14). To successfully conduct psychological operations, one has to know the target population's attitudes and culture. A similar lesson is arguably even truer when it comes to backing foreign political and guerilla groups. As Codevilla argues, "Conducting political warfare by trying to build up foreign groups requires reliable intelligence about those groups' motivations as well as about their capacities. With such intelligence, support of foreign groups is a leap in the dark" (Codevilla, 1989, p. 88). Ultimately, intelligence may not always be able to find the "right" agent (in some cases, such agents might not exist), but at least it can help identify where differences lie.

7. Political warfare can generate unintended consequences.

Political warfare, like all forms of warfare, often comes with unintended consequences. The overthrow of Mossadegh in Iran helped set in motion a series of events that led to the rise of Ayatollah Khomeini. The backing of the mujahedeen in Afghanistan led to the eventual rise of the Taliban several years later. While such kinetic actions provide the most vivid examples of unintended consequences, the same can apply to political and information tools of political warfare as well. The 1976 Church Committee, for example, exposed abuses of power in CIA-funded propaganda campaigns, and those abuses ended up getting published in English and then circulated throughout the American and foreign markets (Turner 1985, p. 82). More recently, the revelations about U.S. intelligence practices by contractor Edward Snowden created worldwide headlines, caused diplomatic tensions, and led Congress to restrict some large-scale data collection and monitoring practices.

Unintended consequences, however, do not always turn out for ill. For example, although the CIA's funding of cultural activities during the Cold War was principally a tool to combat Communism, it had the indirect effect of sponsoring many of the authors, playwrights, and artists who became staples of Western civilization to include playwright Tennessee Williams and historian Arthur Schlesinger, Jr. As University of Michigan literature professor Gorman Beauchamp argues, "A good case could be made . . . that in aiding and abetting these NCL (Non-Communist Left) luminaries, the CIA not only achieved its aim of wooing the Western intelligentsia away from its romance with Communism, but that, however incidentally, it stimulated and enriched the intellectual and artistic life of the West" (Beauchamp, 2001). Needless to say, not all commentators share Beauchamp's positive view.[6] Still, it underscores the fact that political warfare can lead to unpredictable outcomes.

8. Economic leverage is increasingly the preferred tool of the strong.

Over the course of the twentieth century, states have increasingly turned to economic warfare as a tool of political warfare. A recent book by Robert D. Blackwill and Jennifer M. Harris, *War by Other*

6 For a left-wing and generally unfavorable critique of the CIA's funding of cultural activities, see Petras (1999).

Means: Geoeconomics and Statecraft, surveys the rising employment of trade, investment, monetary, commodity, and energy policies as well as economic sanctions, cyberspace, and aid for explicitly political aims. While the United States has in recent decades promoted economic policies on their merits, the authors describe a long history of U.S. economic measures undertaken to increase political advantage. While that approach can backfire if it is seen as a zero-sum game, the authors argue, for example, that an aggressive policy to increase energy independence in Europe and Asia will diminish the leverage that Russia seeks over many Eurasian countries (Blackwill and Harris, 2016, pp. 210, 217). With the growth of globalization and the increasing interconnectivity of world markets, sanctions became an increasingly popular form of coercion. According to data compiled by the Peterson Institute for International Economics, states employed sanctions almost three times as frequently during the 1990s as they did in the first half the twentieth century (see Figure 6.1). As shown in Figure 6.2, the targets of sanctions literally spanned the globe. Between 2000 and 2012, at least 20 states were targeted by sanctions—spread across North America, Africa, Europe, the Middle East, and Asia. Moreover, sanctions prove effective. According to a Peterson Institute study, sanctions produced modest policy changes 51 percent of the time, regime change and democratization 31 percent of the time, a disruption of military adventures 21 percent of the time, and military impairment 31 percent of the time (Hufbauer, Schott, and Elliott, 2008, p. 159).

Despite this dramatic growth, the power of sanctions to coerce remains limited. Perhaps more so than for any other tool of political warfare, size matters when it comes to economic sanctions. Sanctions are a potential tool of the strong—large economies or those endowed with key natural resources—but most other states lack this option. While Russia has tried to turn its oil and natural gas resources into political leverage against Ukraine and Europe, and while the United States has used its economic size to hold sway over Iran, Cuba, and other actors, ISIL and Iran have fewer economic tools at their disposal to influence international politics. They can flout sanctions and use their oil wealth to buy influence, but they lack the ability to sanction.

Figure 6.1
Trends in the Use of Economic Sanctions

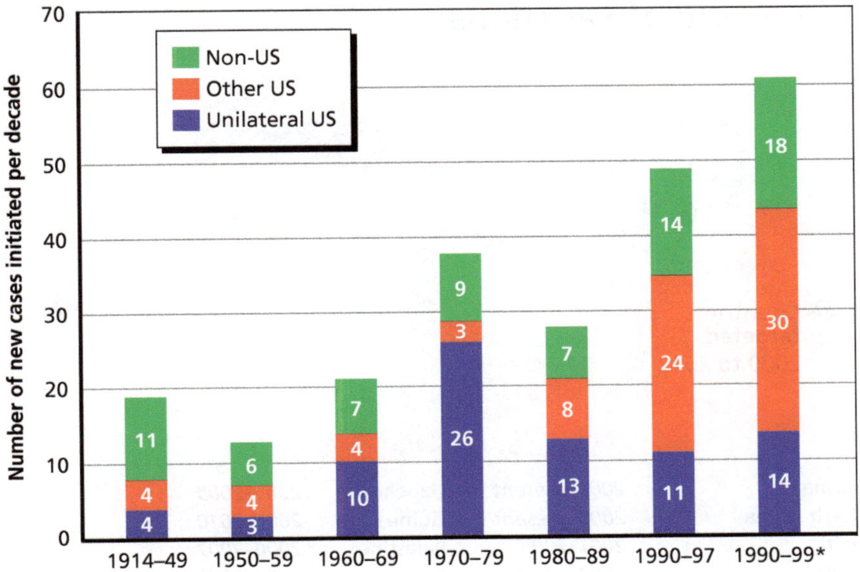

SOURCE: Hufbauer, 1999.
RAND *RR1772-6.1*

Even great powers face constraints. Thanks to the globalized economy, the targets of sanctions can often find ways to mitigate the economic impact of sanctions and lessen their coercive power. For example, Russia provided 39 percent of Europe's natural gas and 33.5 percent of its oil in 2013 (Eurostat, 2015). But when Russia turned off the gas supply to Ukraine after the crisis there in 2013, Europe found workarounds—first by the "reverse flow" of natural gas from Western to Eastern European states and then by slowly shifting to alternative sources of energy (Harrison and Princova, 2015). Even the U.S. record on sanctions is decidedly mixed. Its long-standing sanctions against Cuba failed to prompt that country's transition to democracy. Similarly, sanctions directed against the Democratic Republic of Korea have failed to stop its nuclear program. For a variety of reasons, effective unilateral sanctions often prove challenging (Pape, 1997;

Figure 6.2
Targets of Economic Sanctions

A World of Sanctions

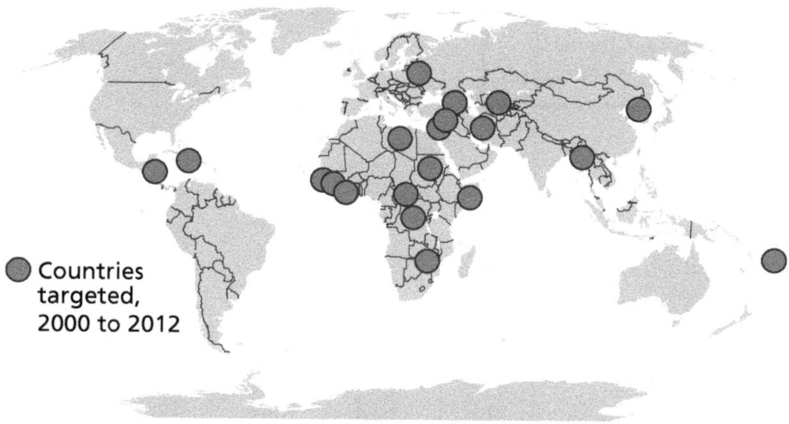

Burma	*2000–present*	Uzbekistan	*2005–2009*
North Korea	*2000–present*	Guinea	*2005–2010*
Haiti	*2001–2005*	Palestine[a]	*2006–2007*
Zimbabwe	*2002–present*	Belarus	*2006–2008, 2010–present*
Dem. Rep. of Congo	*2002–present*	Georgia	*2006–present*
Guinea-Bissau	*2003–2004*	Fiji	*2006–present*
Central African Rep.	*2003–2005*	Honduras	*2009*
Ivory Coast	*2004–present*	Somalia	*2010–present*
Syria	*2004–present*	Iran	*2010–present*
Sudan	*2004–present*	Libya	*2011*

SOURCE: Keane, 2016.
NOTE: Data through May 2012
[a] Hamas-led authority.
RAND *RR1772-6.2*

Pape, 1998; and Elliott, 1998). Sanctions need to be multilateral—to cut off all economic avenues—to be effective, as seen with the international sanctions against Iran to persuade it to halt its nuclear development; even then, though, success is not guaranteed.

Outside of sanctions, actors can use economic inducements as a means of political warfare. States can channel funds to proxies to change the balance of power and to manipulate the internal politics of a targeted state. States can also use the promise of foreign aid to an adversary state to shift its politics. Russia attempted to sway Ukraine's president Viktor Yanukovych away from signing an association agreement with the European Union with the promise of buying $15 billion of Ukrainian bonds and cutting the price of gas from $400 to $268.50 per 1,000 cubic meters.[7] Likewise China is well known for offering foreign aid packages to further its national interests and especially to help it gain access to natural resources.[8]

Liberal democracies also use foreign aid as a means of achieving their political objectives.[9] The United States has used foreign aid to accomplish its political objectives. In the aftermath of the Second World War, the United States embarked on an ambitious endeavor to rebuild Western Europe, partly as a means to forestall the growth of Communism and to check Soviet power. The project blossomed into one of the largest foreign aid endeavors in U.S. history, with over $75 billion (in 2010 dollars) dedicated in a single year. Since then, the American foreign aid budget has remained largely flat, hovering around $25 billion (see Figure 6.3).

Foreign aid—like sanctions—requires significant resources; and even then, success is not guaranteed. Despite its multibillion-dollar offer, Russia's aid to Ukraine failed to prevent its decision to sign an association with the European Union. Still, especially in comparison to the price tag and the risks associated with some other tools of political warfare, wealthy states are willing to take the gamble, making economic warfare an increasingly attractive tool of political warfare for the strong.

---

7 BBC, 2013.

8 For an analysis of China's foreign aid practices, see Wolf, Wang, and Warner, 2013.

9 For general overview, see Alsenia and Dollar, 2000; for an example of United States using foreign aid to influence United Nations Security Council members, see Kuziemko and Werker, 2006; for Japan using foreign aid to influence whaling decisions, see Strand and Tuman, 2012.

shares a common bond with the group, the state may better understand what messages may resonate with the group, what economic incentives might drive it to action, and what military capabilities it might have. Ideally, the state can then turn this understanding into better means for control. Moreover, the closer the shared bond, the more likely the proxy will actually pursue the states'—rather than its own—goals. With a strong enough bond, the two sets of interests become one and the same. In other words, given better knowledge of the proxy, the state might be able to conduct more effective political warfare. In this sense, it should come as no surprise that in both the Russian and Iranian case studies, political warfare efforts worked hand in glove with seemingly benign attempts at cultural promotion—the closer the ethnic or religious bond these groups feel to the motherland, the more easily they are co-opted.

Iran appears to be particularly adept at reaching out to Shi'a constituencies in Iraq, Syria, Lebanon, and Yemen and using aid, trade, religion, education, and a host of "soft power" outreach strategies, in addition to its widely publicized employment of the IRGC, to raise, train, and advise militias in all these countries. Iran has also reached out to Arab audiences the world over with a Pan-Islamic message and has sought to build alliances beyond Shi'a Islam. These overtures are most successful when an intermediary carries the message or burden, as in the case of Iraqi Arab militias sent to largely Arab Syria. Similarly, ISIL relies on indoctrination to forge a common identity among the plethora of foreigners drawn to fight in its ranks.

10. Political warfare extends, rather than replaces, traditional conflict and can achieve effects at lower cost.

Despite its prevalence today, political warfare does not replace conventional warfare but rather extends the spectrum of warfare. Political warfare's tools operate below the threshold of conventional warfare, employing means of coercion short of triggering a full-scale state-on-state conflict. And yet, conventional warfare still exists. In fact, most of the actors analyzed here also fought conventional conflicts. Iran fought Iraq during the 1980s, Russia fought Georgia in 2008 and most recently a limited intervention in Syria. The United States fought wars in Afghanistan, Iraq, and elsewhere. Even ISIL has employed more conventional warfare in Syria and Iraq. Political warfare simply pushes the threshold to lower levels, forming a continuous range of conflict with "no hard-and-fast line" between the more kinetic means of conventional conflict and political warfare (Mazarr, 2015, p. 38).

Defense analyst Michael Mazarr argues that political warfare is unlikely to replace the traditional use of force if nations are willing to take risks to protect their interests:

> Gray zone strategies allow states to capitalize on others' vulnerabilities, but they seldom, if ever, offer avenues to achieve decisive results on their own. Beijing cannot be certain of achieving its ultimate goals in the South China Sea through gradual gray zone tactics and techniques alone. If others resist sufficiently, China will ultimately need to decide whether to escalate to more elaborate forms of aggression. (Mazarr, 2015, p. 121)

What political warfare offers is the chance to achieve an actor's political objectives at a lower cost. If the objectives are sufficiently modest—such as intimidating a weaker adversary or causing destabilization or fragmentation of an alliance—political warfare may suffice. An actor may always opt for political warfare as a first resort and escalate later to hybrid or full-scale conventional warfare if the desired objectives are not met. Sometimes, the actor may settle for a lesser objective if the costs appear to be too high, or if the outcome satisfies current requirements. This could be the case in Georgia, Ukraine, and Moldova, where "frozen conflicts" have stopped the westward leanings of these countries, at least for the moment.

Moreover, as these case studies have demonstrated, political warfare may be most effective in certain economic and social conditions. The Russian, Iranian, and ISIL case studies all illustrate, to varying degrees, that while political warfare may be able to exploit tensions within society or capitalize on existing discontent, these effects are just as much a function of the targets' weakness as of the inherent strength of the aggressor.

For the most part, this study seems to confirm Mazarr's suggestion that political warfare can succeed by itself only in certain circumstances. In most of the case studies, political warfare was used to check a foreign power's influence (as was the case for Iran's use of political warfare against the United States or the U.S. use of political warfare against the Soviet Union) or achieve fairly limited objectives (as with Russia's use of political warfare against Estonia over the Bronze Statue incident). It may achieve larger aims, such as Russia's annexation of Crimea and ISIL's dramatic takeover of territory. These tactics may succeed when no party wishes to escalate or confront the aggressor. This method of advancing aims is best understood as part of the spectrum of warfare. It fills an important niche in the strategic panoply to increase one's influence and leverage while reducing an adversary's influence and leverage, but it does not replace either conventional war or more conventional forces.

One implication that may be drawn from this attribute is that political warfare is a long-term strategy. As Stanfield Turner remarks, "Propaganda changes people's minds only over time; political action usually requires patient, adroit maneuvering; and paramilitary action is guerilla warfare, a gradual wearing down of the enemy over a long period. Even the seeming[ly] quick, decisive, small-scale operation that restored the Shah in 1953 was but the final act of a slow process" (Turner, 1985, p. 88). Turner's point is well taken. Even the early Cold War coups d'état—in Iran, Guatemala, and Chile—did not happen overnight and required lengthy preparation.

As noted above, the elements of political warfare do not work in a vacuum. It usually must be nested in a whole-of-government approach that employs multiple tools—often a balance of force and psychological tools—to accomplish its desired ends. Angelo Codevilla argues that the "Marshall Plan, combined with stationing troops and active black, white and gray propaganda effort, prevents rise of Communism in Western Europe, but it is unlikely that propaganda alone would have been effective" (Codevilla, 1989, p. 84). And while the Marshall Plan needed American military might—and American boots on the ground—to be successful, the converse is also true. Coups d'état in Iran, Guatemala, and Chile required sophisticated political action and information warfare ultimately to succeed.[12] In sum, political warfare is neither simple, nor a silver bullet; success requires a slow, methodical, and multifaceted approach.

Of Terms and Attributes

A heated debate exists today over what to call measures of coercion below the threshold of kinetic conflict. For some, the preferred terms of art are "measures short of war," "irregular warfare," "netwar," "soft war," "hot peace," or simply "operating in the gray zone." Ultimately, as Antulio J. Echevarria argues:

> While the original aim of such labeling, or re-labeling, may have been to draw the attention of busy policymakers to rapidly emerging security issues, it has evolved into something of a culture of replication in which the labels are repeated more out of habit than conscious reflection. This habit has led to a wealth of confusion that has clouded the thinking of policymakers and impaired the development of sound counter-strategies. (Echevarria, 2016, p. 1)

While this monograph has employed the term "political warfare" because of its historical roots, the debate over the correct terminology—while academically important—may be of lesser significance than understanding the phenomenon. In other words, understanding political warfare's attributes may be more important than deciding on the label itself.

The common element in all of these formulations is the application of a host of measures, short of conventional war, as a way to advance an actor's interests without incurring the costs of a full-scale military operation. Indeed, political warfare may share the same objectives as more conventional uses of force—such as weakening or deterring an adversary, changing a regime, or dislodging an occupying power. What varies in political warfare—or the other terms for this

---

12 For example, Codevilla argues that the United States needed to recognize the Shah's restoration in order for the 1953 Iran coups to succeed (Codevilla, 1989, p. 89).

area of activity—is the manner in which these objectives are pursued. Echevarria argues again, that "what makes gray zone conflicts 'interesting' for a contemporary strategist is that they occur below NATO's Article 5 threshold, and below the level of violence necessary to prompt a UN Security Council Resolution" (Echevarria, 2016, p. 12). In other words, political warfare aims to push the boundaries of conflict without crossing an internationally recognized line.

In order to strike this delicate balance, state and nonstate actors need a range of diplomatic/political, informational, and economic measures, as well as uses of military and intelligence forces short of conventional combat. To prevent sparking an accidental conventional war, they frequently obscure attribution for their actions and operate covertly through the use of proxies. In this fight, information warfare plays an important, if double-edged role: Successful manipulation of media can help maintain this shroud of secrecy or at least a fiction of plausible deniability, although thanks to the diversification of media sources, controlling the media becomes an increasingly difficult task. Alternatively, actors can try to prevent political warfare from becoming conventional wars by employing legal means of coercion. For example, sanctions and foreign aid—as tools of economic warfare—have the advantage of being legal and distinct from conflict.

Ultimately, political warfare—both as a term and a concept—itself is not new, but its importance may be increasing. As Mazarr notes, many of these tactics "are classic 'salami-slicing' strategies, fortified with a range of emerging gray area or unconventional techniques—from cyberattacks to information campaigns to energy diplomacy" (Mazarr, 2015, p. 2). But its practice may be on the rise as of late. According to some analysts, political warfare may "be becoming the tool of choice for states wanting to reframe the global order in the 21st century" and may become the "default mode for conflict in the coming decades" (Mazarr, 2015, pp. 4, 101). And so no matter what label one slaps on the concept, it is essential for policymakers to understand its attributes and know how to respond accordingly.

References—Chapter 6

Abdelal, Rawi, "The Profits of Power: Commerce and Realpolitik in Eurasia," *Review of International Political Economy*, Vol. 20, No. 3, June 2013, pp. 421–56.

Abdo, Geneive, Deborah Amos, Reza Aslan, F. Gregory Gause III, Bruce Hoffman, Ed Husain, and Vali Nasr, "The Sunni-Shia Divide," *Council on Foreign Relations*, n.d. As of May 21, 2016:
http://www.cfr.org/peace-conflict-and-human-rights/sunni-shia-divide/p33176#!/?cid=otr-marketing_url-sunni_shia_infoguide

Al Jazeera, "Fact and figures: Global Trade in Small Arms," *Al Jazeera*, March 18, 2013. As of June 1, 2016:
http://www.aljazeera.com/news/americas/2013/03/201331885519413442.html

Alesina, Alberto, and David Dollar, "Who Gives Foreign Aid to Whom and Why?" *Journal of Economic Growth*, Vol. 5, No. 1, March 2000, pp. 33–63.

Arquilla, John, and David Ronfeldt, *Networks and Netwars: The Future of Terror, Crime, and Militancy*, MR1382, Santa Monica, Calif.: RAND Corporation, 2001. As of June 1, 2016:
http://www.rand.org/pubs/monograph_reports/MR1382.html

Bajoria, Jayshree, and Greg Bruno, "Al-Qaeda (a.k.a. al-Qaida, al-Qa'ida)," *Council on Foreign Relations Backgrounder*, June 6, 2012. As of June 30, 2016:
http://www.cfr.org/terrorist-organizations-and-networks/al-qaeda-k-al-qaida-al-qaida/p9126

BBC, "Public Purposes," *Inside the BBC*, n.d. As of April 7, 2016:
http://www.bbc.co.uk/aboutthebbc/insidethebbc/whoweare/publicpurposes/

———, "Russia Offers Ukraine Major Economic Assistance," *BBC News*, December 17, 2013. As of May, 31, 2016:
http://www.bbc.com/news/world-europe-25411118

Beauchamp, Gorman, "Intellectuals Undercover," *Michigan Quarterly Review*, Vol. 40, No. 2, Spring 2001. As of November 24, 2015:
http://quod.lib.umich.edu/cgi/t/text/text=idx?cc=mqr;c=mqr;c=mqarchive;idno=act2080.0040.216;rgn=main;view=text;xc=1;g=mqrg

Black, Ian, "Iran to Launch Spanish-Language Television Channel," *The Guardian*, September 30, 2010. As of May 12, 2016:
http://www.theguardian.com/world/2010/sep/30/iran-spanish-language-television

Blackwill, Robert D., and Jennifer M. Harris, *War by Other Means: Geoconomics and Statecraft*, Cambridge, Mass: The Belknap Press of Harvard University Press, 2016.

Burai, Erna, "Parody as Norm Contestation: Russian Normative Justifications in Georgia and Ukraine and Their Implications for Global Norms," *Global Society*, Vol. 30, No. 1, 2016, pp. 67–77.

Codevilla, Angelo M., "Political Warfare," in *Political Warfare and Psychological Operations: Rethinking the US Approach*, Carnes Lord and Frank R. Barnett, eds., Washington, D.C.: National Defense University Press, 1989.

Cohen, Raphael S., "Putting a Human and Historical Face on Intelligence Contracting," *Orbis*, Vol. 54, No. 2, Spring 2010, pp. 232–51.

Cole, J. Michael,"Unstoppable: China's Secret Plan to Subvert Taiwan," *National Interest*, March 23, 2015. As of April 5, 2016:
http://nationalinterest.org/feature/unstoppable-chinas-secret-plan-subvert-taiwan-12463?page=2

Department of the Treasury, *National Terrorist Financing Risk Assessment*, 2015. As of June 1, 2016:
https://www.treasury.gov/resource-center/terrorist-illicit-finance/Documents/National%20Terrorist%20Financing%20Risk%20Assessment%20%E2%80%93%2006-12-2015.pdf

Dunlap, Jr., Charles J., "Lawfare: A Decisive Element of 21st-Century Conflicts?" *Joint Forces Quarterly*, Vol. 54, No. 3, 2009, pp. 34–39. As of May 13, 2016:
http://www.dtic.mil/dtic/tr/fulltext/u2/a515192.pdf

Echevarria II, Antulio J., *Operating in the Gray Zone: An Alternative Paradigm for U.S. Military Strategy*, Carlisle Barracks, Pa.: Strategic Studies Institute, United States Army War College Press, 2016.

Elliott, Kimberly A., "The Sanctions Glass: Half Full or Completely Empty?" *International Security*, Vol. 23, No. 1, Summer 1998, pp. 50–65.

Eurostat, "Energy Production and Imports," *Statistics Explained*, 2015. As of May 31, 2016:
http://ec.europa.eu/eurostat/statistics-explained/index.php/Energy_production_and_imports

Finnemore, Martha, "Constructing Norms of Humanitarian Intervention," in *The Culture of National Security*, ed. Peter Katzenstein, New York: Columbia University Press, 1996.

Freedom House, "Russia Targets Key Media and Journalists in Crimea," *Freedom House*, March 4, 2014. As of May 13, 2016:
https://freedomhouse.org/article/russia-targets-key-media-and-journalists-crimea

Giles, Keir, *Russia's 'New' Tools for Confronting the West Continuity and Innovation in Moscow's Exercise of Power*, London: Chatham House, March 2016. As of May 21, 2016:
https://www.chathamhouse.org/sites/files/chathamhouse/publications/research/2016-03-21-russias-new-tools-giles.pdf

Graves, Lucas, "Everyone's a Reporter," *Wired Magazine*, September 1, 2005. As of May 13, 2016:
http://www.wired.com/2005/09/everyones-a-reporter/

Haass, Richard N. "The Age of Nonpolarity: What Will Follow U.S. Dominance?" *Foreign Affairs* (May/June 2008): 44–56.

Harrison, Colin, and Zuzana Princova, "A Quiet Gas Revolution in Central and Eastern Europe," *Energy Post*, October 29, 2015. As of June 1, 2016: http://www.energypost.eu/quiet-revolution-central-eastern-european-gas-market/

Hashem, Ali, "Hezbollah's Media Machine Changes Gears," *Al-Monitor*, June 5, 2015. As of May 13, 2016: http://www.al-monitor.com/pulse/originals/2015/06/hezbollah-media-manar-changing-syria-israel-message.html#

Hobson, Peter, "Calculating the Cost of Russia's War in Syria," *The Moscow Times*, October 20, 2015. As of May 21, 2016: http://www.themoscowtimes.com/business/article/calculating-the-cost-of-russias-war-in-syria/540015.html

Hoffman, Bruce, *Inside Terrorism*. New York: Columbia University Press, 2006.

"How NATO's Article 5 Works," *Economist*, March 9, 2015. As of May 13, 2016: http://www.economist.com/blogs/economist-explains/2015/03/economist-explains-6

Hufbauer, Gary Clyde, "Trade as a Weapon," Fred J. Hansen Institute for World Peace, San Diego State University, April 12, 1999. As of April 18, 2016: https://piie.com/commentary/speeches-papers/trade-weapon

Hufbauer, Gary Clyde, Jeffrey J. Schott, and Kimberly Ann Elliott, *Economic Sanctions Reconsidered*, Petersen Institute, 2008. As of April 21, 2016: https://piie.com/publications/chapters_preview/4075/06iie4075.pdf

Ikle, Fred C., "The Modern Context," in *Political Warfare and Psychological Operations: Rethinking the US Approach*, Carnes Lord and Frank R. Barnett, eds., Washington, D.C.: National Defense University Press, 1989.

"Iran Quds Chief Visited Russia Despite U.N. Travel Ban," Reuters, August 7, 2015. As of January 23, 2017: http://www.reuters.com/article/us-russia-iran-soleimani-idUSKCN0QC1KM20150807#THalTZdtqtlu3QeK.97.

Jurkowitz, Mark, "The Growth in Digital Reporting: What It Means for Journalism and News Consumers," *Pew Research Center*, March 27, 2014. As of May 13, 2016: http://www.journalism.org/2014/03/26/the-growth-in-digital-reporting/

Kaufman, Stuart J., *Modern Hatreds: The Symbolic Politics of Ethnic War*, Ithaca N.Y.: Cornell University Press, 2001.

Keane, Angela Greiling, "Sanctions: Financial Warfare," *Bloomberg*, January 19, 2016. As of April 18, 2016: http://www.bloombergview.com/quicktake/financial-war

Kitson, Frank, *Low Intensity Operations: Subversions, Insurgency, and Peacekeeping*, London: Faber and Faber, 1973.

Krasnoboka, Natalya, "Media Landscapes: Russia," *European Journalism Center*, n.d. As of May 21, 2016:
http://ejc.net/media_landscapes/russia

Kuziemko, Ilyana, and Eric Werker, "How Much Is a Seat on the Security Council Worth? Foreign Aid and Bribery at the United Nations," *Journal of Political Economy*, Vol. 114, No. 5, October 2006, pp. 905–30

Lord, Carnes, "The Psychological Dimension in National Strategy," in *Political Warfare and Psychological Operations: Rethinking the US Approach*, Carnes Lord and Frank R. Barnett, eds., Washington, D.C.: National Defense University Press, 1989.

Masters, Jonathan, and Zacharry Laub, "Hezbollah (a.k.a. Hizbollah, Hizbu'llah)," *CFR Backgrounder*, January 3, 2014. As of June 30, 2016:
http://www.cfr.org/lebanon/hezbollah-k-hizbollah-hizbullah/p9155

Mazarr, Michael J., *Mastering the Gray Zone: Understanding a Changing Era of Conflict*. Carlisle Barracks, Pa.: Strategic Studies Institute, United States Army War College Press, 2015.

Morganthua, Hans J., *Politics Among Nations: The Struggle for Power and Peace*, 7th ed., New York: McGraw Hill, 2006.

McDermott, Roger, "Russia Activates New Defense Management Center," *Eurasia Daily Monitor*, Vol. 11, No. 196, November 4, 2014. As of January 18, 2017:
http://www.jamestown.org/programs/edm/single/?tx_ttnews%5Btt_news%5D=43041&cHash=60f76b139c88bc8177cb5cba105c4fe6#.VrkNYhFUODE

Mueller, John, "The Banality of 'Ethnic War,'" *International Security*, Vol. 25, No. 1, Summer 2000, pp. 42–70.

Olmstead, Kenneth, and Elisa Shearer, "Digital News—Audience: Fact Sheet," *Pew Research Center*, April 29, 2015. As of May 13, 2016:
http://www.journalism.org/2015/04/29/digital-news-audience-fact-sheet/

Omang, Joanne, "A Historical Background to the CIA's Nicaragua Manual," in *Psychological Operations in Guerilla Warfare: The CIA's Nicaragua Manual*. New York: Vintage Books, 1985.

Overy, Richard, "Making and Breaking Morale: British Political Warfare and Bomber Command in the Second World War," *Twentieth Century British History*, Vol. 26, No. 3, July 2015, pp. 370–99

Pahlavi, Pierre, "Understanding Iran's Media Diplomacy," *Israel Journal of Foreign Affairs*, Vol. 6, No. 2, 2012, pp. 21–33.

Pape, Robert, "Why Economic Sanctions Do Not Work," *International Security* 22, No. 2, Fall 1997, pp. 90–136

———, "Why Sanctions Still Do Not Work," *International Security*, Vol. 23, No. 1, Summer 1998, pp. 66–77.

Parameswaran, Prashanth, "Beware China's Political Warfare Campaign Against US, Allies: Experts," *The Diplomat*, October 23, 2015. As of November 30, 2015: http://thediplomat.com/2015/10/beware-chinas-political-warfare-campaign-against-us-allies-experts/

PBS, "World War II Propaganda," *American Experience*, n.d. As of April 7, 2016: http://www.pbs.org/wgbh/amex/goebbels/peopleevents/e_propaganda.html

Peters, Rebecca, "Small Arms: No Single Solution," *UN Chronicle*, Vol. 4, Nos. 1–2, 2009. As of June 1, 2016: http://unchronicle.un.org/article/small-arms-no-single-solution/

Petras, James, "The CIA and the Cultural Cold War Revisited," *Monthly Review*, Vol. 51, No. 6, November 1999. As of November 24, 2015: http://monthlyreview.org/1999/11/01/the-cia-and-the-cultural-cold-war-revisited/

"Progress," USAID, February 16, 2016. As of April 19, 2016: https://www.usaid.gov/results-and-data/progress

Raska, Michael, "China and the 'Three Warfares,'" *The Diplomat*, December 18, 2015. As of April 5, 2016: http://thediplomat.com/2015/12/hybrid-warfare-with-chinese-characteristics-2/

"Revolutionary Armed Forces of Colombia—People's Army," *Mapping Militant Organizations*, Stanford University, August 15, 2015. As of June 30, 2016: http://web.stanford.edu/group/mappingmilitants/cgi-bin/groups/view/89

Ria Novosti, "Istorija agentstva," [RIA News, "Agency History"], n.d. As of April 24, 2016: http://ria.ru/docs/about/ria_history.html#ixzz3miMy1Gng

Rosenberg, Steven, "Ukraine Crisis: Meeting the Little Green Men," *BBC News*, April 30, 2014. As of May 13, 2016: http://www.bbc.com/news/world-europe-27231649

RT.com, "More Money, More Problems?" February 19, 2015. As of January 18, 2017: https://www.rt.com/op-edge/233147-media-rt-financing-myths-facts/

Salehyan, Idean, "The Delegation of War to Rebel Organizations," *Journal of Conflict Resolution* 54, No. 3, June 2010, pp. 493–515.

Schaefer, Brett D., and Anthony B. Kim, "The U.S. Should Link Foreign Aid and U.N. General Assembly Voting," *Heritage Foundation Backgrounder*, August 8, 2011. As of April 18, 2016: http://www.heritage.org/research/reports/2011/08/the-us-should-link-foreign-aid-and-un-general-assembly-voting

Scott, Len, "Secret Intelligence, Covert Action and Clandestine Diplomacy," *Intelligence and National Security*, Vol. 19, No. 2, Summer 2004, pp. 322–41.

Shevchenko, Vitaly, "'Little Green Men' or 'Russian Invaders?'" *BBC News*, March 11, 2014. As of May 13, 2016:
http://www.bbc.com/news/world-europe-26532154

Shorrock, Tim, *Spies for Hire: The Secret World of Intelligence Outsourcing*, New York: Simon and Schuster, 2009.

Shuster, Simon, "Inside Putin's On-Air Machine," *Time*, March 16, 2015, pp. 44–51.

Singer, Peter, *Corporate Warriors: The Rise of the Privatized Military Industry*, Ithaca, N.Y.: Cornell University Press, 2007.

Smyth, Phillip, "Iran-Backed Iraqi Militias Are Pouring into Syria," *Business Insider*, October 2, 2015. As of January 20, 2017:
http://www.businessinsider.com/
iran-backed-iraqi-militias-are-pouring-into-syria-2015-10

Speier, Hans, "The Future of Psychological Warfare," *Public Opinion Quarterly*, Spring 1948, pp. 5–18.

Stokes, Mark, and Russell Hsiao, "The People's Liberation Army General Political Department: Political Warfare with Chinese Characteristics," *Project 2049*, October 14, 2013. As of April 5, 2016:
http://www.project2049.net/documents/PLA_General_Political_Department_
Liaison_Stokes_Hsiao.pdf

Strand, Jonathan R., and John P. Tuman, "Foreign Aid and Voting Behavior in an International Organization: The Case of Japan and the International Whaling Commission," *Foreign Policy Analysis*, Vol. 8, No. 4, October 2012, pp., 409–30.

Szczepański, Marcin, "Economic Impact on the EU of Sanctions over Ukraine Conflict," *European Parliamentary Research Service*, PE 569.020, October 2015. As of May 21, 2016:
http://www.europarl.europa.eu/RegData/etudes/BRIE/2015/569020/
EPRS_BRI(2015)569020_EN.pdf

Tarnoff, Curt, and Marian Leonardo Lawson, "Foreign Aid: An Introduction to U.S. Programs and Policy," *Congressional Research Service,* R40213, February 10, 2011. As of April 18, 2016:
https://www.fas.org/sgp/crs/row/R40213.pdf

Tétrault-Farber, Gabrielle, "Looking West, Russia Beefs Up Spending on Global Media Giants," September 23, 2014. As of May 12, 2016:
http://www.themoscowtimes.com/news/article/looking-west-russia-beefs-up-spending-on-global-media-giants/507692.html

Tsagourias, Nicholas, "Cyber Attacks, Self-Defence and the Problem of Attribution," *Journal of Conflict & Security Law*, July 2012, pp. 1–16.

Turner, Stanfield, *Secrecy and Democracy: The CIA in Transition*, Boston: Houghton Mifflin Company, 1985.

U.S. Agency for International Development, *U.S. Overseas Loans and Grants* (Greenbook), Washington, D.C., multiple years.

Wolf, Jr., Charles, Xiao Wang, Eric Warner, *China's Foreign Aid and Government-Sponsored Investment Activities: Scale, Content, Destinations, and Implications*, Santa Monica, Calif.: RAND Corporation, RR-118, 2013. As of May 31, 2016: http://www.rand.org/pubs/research_reports/RR118.html

CHAPTER SEVEN

Influence Communications: Lessons for the United States

Information warfare is a critical component of political warfare. Indeed, information activities play a prominent role in each of the case studies presented in this report. During the Cold War, the Voice of America broadcast programs in over 30 languages. Radio Free Europe and Radio Liberty and other U.S. programs used leaflet-laden balloons and covertly funded cultural activities to counter the Soviet threat. While U.S. informational activities continue, the United States Information Agency was disbanded after the Cold War. Today, Russia, Iran, and ISIL devote substantial funding and organizational manpower to conduct messaging. Russia runs two major satellite television agencies to broadcast pro-Russia content both within Russia and around the world. Russia also makes effective use of social media by using bots and trolls on Twitter to support Russian propaganda themes and to attack adversaries. Iran not only broadcasts propaganda on satellite television but also uses an array of proxies and cultural programming to convey its message. For its part, ISIL has revolutionized the art of propaganda by enlisting its cadre of supporters around the world to help disseminate ISIS-produced content on social media.

Given the importance that state and nonstate actors place on information warfare, it seemed prudent for this study to examine modern-day U.S. communication efforts.[1] Consequently, we interviewed over

[1] We should note that the focus of this analysis is on U.S. Government communication, information and influence capabilities as opposed to strategic communications writ large. According to Joint Publication 1-02, Department of Defense Dictionary of Military and

20 active and retired U.S. government experts in persuasive messaging. In accordance with our scope, the vast majority of these experts worked within the U.S. Department of State (DoS) and U.S. Department of Defense (DoD), and they shared a specific focus on both the newly developed Global Engagement Center (GEC) within DoS and the MISO forces within DoD.[2] During these interviews, we asked participants to identify key challenges to U.S. communication efforts and ways the United States could address these challenges.

These interviews yielded an abundance of information, out of which we identified six critical themes. These themes include the following:

- Leadership must empower communicators to quicken response times and take risks.
- The pros and cons of non-attribution need to be weighed.
- The use of influencers is a critical requirement for both the GEC and MISO.
- MISO requires both increased manpower and new media training.
- CENTCOM's model of web operations could be emulated.
- Interagency coordination remains a fundamental challenge.

Associated Terms (Washington, D.C.: The Joint Staff, November 10, 2010), the strategic communications is defined as "Focused United States Government efforts to understand and engage key audiences to create, strengthen, or preserve conditions favorable for the advancement of United States Government interests, policies, and objectives through the use of coordinated programs, plans, themes, messages, and products synchronized with the actions of all instruments of national power." Christopher Paul of RAND provides an excellent critique of this term (Paul, 2010) in which he argues that the term encompasses a panopoly of meanings to include (1) "Enterprise level strategic communication"; (2) "Strategic communication planning, integratin and synchronization processes"; (3) "Communication strategies and themes"; (4) "Communication, information and influence capabilities"; and (5) "Knowledge of human dynamics and analysis or assessment capabilities." For this reason we specifically scoped this chapter to focus on communication, information and influence capabilities.

2 We recognize the limitations of this study; for instance, we did not interview representatives of the Broadcasting Board of Governors, public affairs officers within the Department of Defense, representatives of the National Security Council, or U.S. commanders overseeing operations in Iraq.

We should note that several key organizational developments in the civilian U.S. government information bureaucracy inform this analysis.[3] First, as noted, the USIA was dismantled in 1999. Originally established in 1959, the USIA served as an independent agency under the executive branch charged with conducting public diplomacy in support of U.S. foreign policy. Its mission was to explain and advocate U.S. policies abroad, provide information about U.S. policy and the U.S. people, facilitating strong connections between Americans at home and those abroad and to advise the President and U.S. government on the relationship between U.S. policy and foreign attitudes (USIA Overview, 1998). The organization led the propaganda fight against the Soviet Union and was then dismantled in 1999. Its communication portfolio was split up between the U.S. State Department and the newly created Broadcasting Board of Governors which now controls the Voice of America, Radio Free Europe/Radio Liberty, Radio Free Asia, Radio and Tv Martí, and the Middle East Broadcasting Networks (BBG, 2016).

Starting in 2011, the U.S. Department of State underwent two consecutive organizational changes to enhance its ability to counter-message al Qaeda and subsequently ISIL. First, to counter al Qaeda's media and recruitment operations, the Obama administration established the Center for Strategic Counterterrorism Communications (CSCC). The CSCC was charged in part with "using communication tools to reduce radicalization by terrorists and extremist violence and terrorism that threaten the interests and national security of the United States" (White House, 2011). Second in March 17, 2016, the White House issued Executive Order 13721, which effectively dissolved the CSCC and in its place established the Global Engagement Center (GEC) within DoS (White House, 2016). The GEC mission is two-fold. First, it has as a key responsibility for coordination and synchronization of U.S. government communication efforts against the Islamic

---

3 Of course there has also been development on the Department of Defense sector to include the re-naming of the military's Psychological Operations forces, Military Information Support Operations Forces, the development of web capabilities at U.S. Central Command (see below section titled, "CENTCOM's model of web operations could be emulated").

State.[4] Second, rather than serve as a central hub for U.S. government–directed communications, the GEC seeks to help partner governments and civil society actors to create and disseminate their own content in support of U.S. counter-terrorism objectives (White House, 2016).

1. Leadership must empower communicators to quicken response times and take risks.

Rapid communications are inherently risky. The news media seem to devote an extraordinary level of attention to government communication efforts. Within DoS, the Global Engagement Center (GEC) and its DoS predecessor, the Center for Strategic Counterterrorism Communications (CSCC), have both been subject to numerous media exposés (see for example, Cooper, 2016; Miller, 2015; Silverman, 2016).[5] As one senior official from the CSCC noted, for a bureaucratically small and underfunded organization, the CSCC had received a "huge amount of attention because of what it was supposed to do" (interview 2, 2016).

With this attention comes bureaucratic risk. The DoD Office of Strategic Influence, developed shortly after 9/11 to counter jihadi ideology, was shuttered due to publicity (Hess, 2002). Various sources suggest that a segment by John Oliver on his HBO talk show *Last Week Tonight* that ridiculed a CSCC video, "Welcome to ISIS Land," played a critical role in hastening the office's demise (interviews 2 & 10, 2016). A former CSCC official observed that the White House will "hang you out to dry. . . . They will quickly tout your successes, but if something goes wrong, they will not back you up" (interview 5, 2016). Just speaking to the press carries risk. "Sometimes, people have a wide discretion, but you are taking your career in your hands if you speak to the press

[4] The Executive Order establishing the CSCC also called for interagency coordination in counter-terrorism communications.

[5] Per Executive Order 13721, the GEC is charged with supporting government-wide counter-terrorism communication efforts that are directed at foreign audiences (White House, 2016). The GEC replaced the CSCC, an office created on September 9, 2001.

without clear guidance. . . . If you say the wrong thing, you are toast; nobody protects you" (interview 6, 2016).

This risk of publicity and the consequent fear of losing one's career have several critical implications for communications. First and most significantly, communications are slow. Within DoS, the process for answering media questions is a measured and deliberate one with great care taken to ensure that relevant embassies and representatives of the regional and functional bureaus, and sometimes even the White House, have an opportunity to clear answers to anticipated policy questions. Even at the CSCC, which has often engaged in tweet wars with ISIS, responses are relatively slow. A former CSCC official lamented that the "hierarchical nature of government makes it mis-aligned and inflexible when it comes to responding quickly." In contrast to the "real world" where you would respond to a tweet within seconds or minutes, they would at best respond within hours. "That is fast for the U.S. Government. It is slower than the real world," observed the official.[6] Many noted that communications were also slow in the military. According to doctrine, the combatant commander has authority to approve individual influence messages for dissemination but that the commander can delegate the decision down to a subordinate level. One MISO officer commented that many commanders are very risk-averse and maintain approval at the general officer level, a decision certain to delay messaging, especially in the new media environment (interview 5, 2016). In addition, relevant U.S. ambassadors must also approve messages disseminated within their countries. As one bright spot, however, web operations within CENTCOM benefit from a streamlined approval process in which messaging themes are pre-approved by DoS, while military approval authority rests below the rank of colonel. This allows units to respond quickly to developing events and adversary messages (interview 11, 2016).

---

6 The challenge apparently became worse when CSCC disseminated content in English. "Once we went into English, we got into trouble and we would respond too quickly. In U.S. government terms, we were responding too fast, we were reckless, you see. Slower than the real world and too fast for the government world. Government is about control" (interview 2, 2016).

Beyond speed, the second major challenge with bureaucratic communications is risk-taking. This is especially felt within the GEC, where, as a high-profile and newly standing organization, staffers feel that a wrong or inadvertent turn can jeopardize the organization's bureaucratic life. Some staffers, however, noted the importance of failure in learning and ultimately improving operations. The technology sector learns by failure, they note, and ISIS does, too. ISIS can disseminate "ten IO themes, and maybe one of them sticks. We can't do that. We are so afraid that it will end our funding" (interview 10, 2016). Several staffers walked through different concepts that they felt were necessary to push the envelope against ISIS and to help restrain the large number of potentially radicalizing individuals that can be identified on public social media channels. But in the words of one analyst, "Everyone is terrified to try."

Related to this reticence is a general fear of countering ISIS ideology using the English language. According to a former CSCC official,

> Once it is in English, it is everyone's business. Everyone is a critic, everyone has an opinion. The traffic in this space is excessive. In Arabic, you don't have to worry about the *New York Times* looking over your shoulder or John Oliver or ACLU. But once in English, you are part of the English language twitterverse. [We get it] from the right and from the left. We get it from Islamists and anti-Islamists. If you go in English, you are taking your life in your hands. (Interview 2, 2016)

Currently, GEC's primary focus is on its Arabic-speaking audience in the Middle East and so communication in English is not a priority. However, concerns persist that communicating in English is a dangerous and best-avoided endeavor. As one interlocutor noted, it is best to be "less turbulent in the public world" (interview 10, 2016).

Implications

Success in counter-messaging requires senior leaders, from the White House to military commanders, to empower subordinates to accept risk. In reference to influence communications, a former CSCC official argued that communicators "need to be linked to political leadership and empowered. . . . You need an entity that will have political cover from the White House."

This has several implications. First, communicators should develop systems that allow them to quicken the pace of messaging approval and response. CENTCOM web operations are a prime example of effectively using DoS coordination, fast review cycles, and subordinate delegation to expedite response times to at least approximate the speed of the World Wide Web. To be effective, senior leaders must trust more junior subordinates to clear products within appropriate limits.

Second, senior leaders must not just empower but also *encourage* subordinates to accept risk. With the development of effective and speedy processes for approving products, it is critical that communicators take reasonable risks in the pursuit of success. This will include using the English language to counter U.S. adversaries. With this risk-taking, mistakes will happen; and a press, fascinated by propaganda, will take notice. Senior leaders, especially within the White House, must be willing to provide the necessary political cover to the bureaucracy, modify plans and operations accordingly, and push on.

2. Unattributed communications may have counterproductive effects that should be anticipated and mitigated.

One question that confronts military influence planners is whether or not to attribute message products directly to the U.S. government. According to doctrine, the combatant commander can choose to directly acknowledge U.S. created products, acknowledge in a delayed fashion for sensitive activities, or attribute the product to a concurring partner nation. But the DoS does not have such choices and can disseminate products only with U.S. acknowledgment.

Interviews with MISO officers and staff suggest different arguments for and against direct attribution. The primary argument in favor of non-attribution focuses on credibility. Specifically, the United States lacks credibility in some regions of the world and among some specific target audiences. In such cases, MISO efforts can have a greater effect if they are unattributed or if their attribution is delayed until a later point in time. Observed one MISO staffer at the U.S. Special Operations Command (SOCOM), "Target audiences are most receptive when they trust the source. How they feel about the source is important. And with U.S. government attribution, that trust is reduced. So non-attribution is [often the] best." Another observed, "There are some target audiences that can't be reached with an attributed voice" (interview 1, 2016).

In contrast, attributed messages have their advantages. First, using non-attributed communications is a bureaucratically cumbersome process, requiring a specific legal finding and senior official approval. One strategic communications expert noted that one can often be "more agile and responsive" by using U.S. attribution. "We would rather have an effect immediately with attributed voice than get a traditional military activity finding" (interview 1, 2016). In addition, he argues, the need for non-attribution really depends on the message and target audience. "I think there are a lot of effects you can have with attributed voice." The downside of routinely going with a non-attributed voice "is you miss the realm of the possible with an attributed voice. Ways you can mobilize others, shape opinions, and get messages out" (interview 1, 2016). Non-attributed messages lack the authenticity and personality that otherwise come with proxy influencers who have their own history, personality, and following. Non-attribution also carries its own political risks. During Operation Iraqi Freedom (OIF), one influence program called for the creation of non-attributed video games targeted at Middle Eastern teenagers in "high risk/unfriendly areas." The game provided players a first-person shooter scenario where they could play the role of the "Iraqi Hero" who shot insurgents trying to kill civilians. Ultimately, the program was shut down in part due to negative publicity over its non-attribution status (interview 5, 2016).

3. The use of third-party influencers is a critical requirement for both the GEC and MISO.

As previously noted, on March 17, 2016, the GEC replaced the CSCC as the main hub for counter-terrorism communications. The CSCC employed a variety of tactics to counter ISIS but ultimately was maligned for its attempts to directly tweet against ISIS using a U.S.-attributed Twitter account (Suebsaeng, 2014). The GEC has subsequently sought to limit such a direct and "top-down" messaging effort and instead focuses on helping both government and civil society partners to develop and disseminate counter-extremism messages. Its director Michael Lumpkin noted, that "this 'partners-first' approach is necessary to change the context and stimulate and amplify the voices of influential and peace-promoting actors to counter the violent extremist propaganda in their regions, and expose Daesh's [ISIL's] true nature."

The use of this partners approach theoretically offers several key advantages over alternative U.S.-centric messaging efforts. First, in many ways it mimics ISIS's own messaging strategy, which relies on a cadre of supporters as well as key elites to send credible messages and amplify ISIL's own propaganda messages. In addition, the approach avoids a critical weakness of U.S.-centric campaigns, specifically that the United States is not seen as a credible source among those in the Middle East and elsewhere who are most at risk of radicalization. Second, it overcomes a U.S. reluctance to directly counter the theological underpinnings of radical Islamic ideology, owing to the particular lack of U.S. credibility in interpreting Islamic theology and concerns about separation of church and state. Third-party anti-ISIS influencers in the Middle East are under no limitation and can use theological arguments, drawing on Koranic texts, to undermine the ISIS message.

To accomplish this objective, the GEC works to empower a variety of different partners. It seeks to promote engagement from various state actors, to include those in the Middle East, to more actively counteract ISIS messages. An example of this is the Sawab Messaging Center, a counter-ISIS online messaging center that is a joint U.S.-UAE effort (State Department, 2015). A similar center is under development in Malaysia. The GEC also seeks to work with civil society and nongovernmental organizations to promote the creation of content and raise the profile of key and local, credible anti-extremism messengers (interview 9, 2016). In one example, the GEC is working to establish an online radio station in east Africa that will air youth-produced programming that seeks to counter the rise of extremist propaganda in the region (Lumpkin, 2016). As Richard A. Stengel, Under Secretary for Public Diplomacy and Public Affairs, noted in testimony before Congress, "Instead of direct messaging to potential ISIL sympathizers, much of our work focuses on supporting and empowering a global network of partners—from NGOs to foreign governments to religious leaders—who can act as more credible messengers to target audiences" (Stengel, 2016).

Of course, such efforts face many challenges. Extremists have sought to silence such influencers with threats and intimidation. ISIS recently issued death threats against several prominent Muslim theological actors (Goodstein, 2016). Some have commented that there is a limited resource of effective and mature civil society actors in the Middle East (interview 10, 2016). Even willing, credible voices must receive some form of financial incentive to spend their time and resources developing and disseminating content. The GEC also experiences a number of bureaucratic challenges operating within the Department of State.[7] While the ultimate success of this approach is an open question, the GEC seems aware of and seeks to address many of these challenges.

The DoD undertakes its own approach to working with influencers. During Operation Iraqi Freedom, psychological operations units promoted events that highlighted prominent Iraqi artists who called for peace and democratic governance. Today in Iraq, MISO units are performing an advise-and-assist role to help Iraqi forces improve their

---

[7] The GEC is positioned within DoS, and this position has been a source of friction for both GEC and DoS officials. Conversations with the start-up GEC and traditional state department staffers suggest a continuing friction between the organizations. GEC staffers report being encumbered by State's byzantine bureaucratic offices. One GEC staffer noted, "We are value added to our partners if we are quick and agile. But at State we are 'slowed to the speed of the building" (interview 10, 2016). Part of the challenge is that the GEC does not have grant authority, a key weakness given that part of their mission is to provide funding to third-party influencers. Obtaining such grants today requires that GEC staffers work closely with regional bureaus. The staffers must also clear GEC field programs with individual country ambassadors, a process that can significantly delay and stifle communications.

Within the broader State Department, there is a perception among at least a few officials that the GEC could be more cooperative with the broader State Bureaucracy (interview 6, 2016). It was argued for example that the organization had an independent streak that did not open doors in a State Department that cherishes relationships and cooperation (interview 6, 2016).

own influence efforts (Moore, 2015). MISO units are also working to empower anti-ISIS voices on the Internet with sharable anti-ISIS content (see section below on CENTCOM web ops) (interview 11, 2016). In addition, one prominent MISO officer, as part of his United States Army War College thesis, strongly argued for a strategic influence approach centered on supporting "indigenous natural allies" to "create a competing social movement and ideological narrative" (Garcia, 2013).

Despite these efforts, MISO doctrine is relatively quiet on the topic of influencers. Joint Publication 3-13.2 Military Information Support Operations, FM 3-05.301 (2007) Psychological Operations Process Tactics, Techniques and Procedures and FM 3-53 (2013) Psychological Operations, key doctrine texts for MISO, remain relatively silent on the issue. JP 3-13.2, for example, devotes most of its attention on this issue as part of a reference to a brief paragraph on "indigenous assets." Most of this and the other documents, however, focus on U.S. focused product development and dissemination.

Implications

The use of proxies, though more challenging than conventional top-down communications, represents a potentially effective approach to overcoming limited U.S. government credibility while also harnessing the influence, following, and authenticity of local credible actors. We recommend that MISO forces as well as the GEC and other communication elements within the U.S. government increasingly look for opportunities to work with relevant proxies as a way to supplement other communication campaigns. This would include working MISO forces working with relevant civil society actors in locations allowed by current authorities as well as focusing on building the capacity of partner nation psychological operations forces. In particular, this should be a goal of the SOCOM working group currently studying future employment of digital communications. Developing doctrine to guide such efforts will be increasingly critical.

4. MISO requires both increased manpower and new media training.

MISO is challenged by significant manpower shortages and limited new media training. Numerous interviewees spoke about manpower shortages for MISO units. According to one senior MISO officer, "There is a huge demand signal from operations. We are operating on a 1:1 cycle [time spent in garrison vs. in deployment] when SOCOM is pushing for a 2:1 cycle. There is demand for what we do, and that has reached a level not seen before" (interview 8, 2016). With this 1:1 cycle, a number of potentially unmet needs could be addressed with more manpower, the interviews suggest. First, currently only ten O-6 MISO billets exist, and several interviewees suggested that more O-6 personnel could help fill several strategic positions, such as an "integration and deconfliction officer" for MISO teams assigned to U.S. embassies in Europe. Meanwhile, interviews with the GEC suggested that it could also benefit from additional lower-grade MISO officers and enlisted personnel who could help fill in a major staffing deficit within GEC (interview 10, 2016). Third, some organizations such as CENTCOM have had to resort to hiring contractors because of too few trained MISO soldiers. Fourth, it was argued that it is critical to place MISO personnel in Pentagon and Combatant Command (COCOM) plans in order to help military planners "bake in" the messaging and influence operations into their military planning efforts. Regardless of the merit of each of these individual proposals, it does appear that there is growing demand for MISO personnel.

Beyond manpower shortages, others attest to training limitations. This training deficit is most acutely noticed with regard to web operations. Web influence operations in CENTCOM currently rely on contracted support to fill in key communications and language specialty positions. In addition, a senior MISO officer noted, "We need to educate the force, and pipeline does not do that. . . . How do we deal with Facebook, Snapchat, and Instagram? How do we leverage deep data analytics" (interview 8, 2016)? This was also the focus of testimony in front of the House Armed Services Committee by Major General Christopher Haas, U.S. Special Operations Command

Director, Directorate of Force Management and Development. He noted, "The current conflicts have identified that we have a need to continue expanding our MISO training, primarily with regards to the Web" (Haas, 2015).

As also noted by MG Haas, SOCOM is in the process of developing a comprehensive plan capturing all aspects of this requirement—a key aspect of the requirement being training.

> This training will incorporate social media use, online advertising, web metrics, and web design, among many other topics. Such a training solution will also enable us to stop being so dependent upon a contracted solution. In the interest of managing expectations, however, such training cannot happen overnight, and we may always need some level of contracted support in translation and IT expertise. We will be dependent upon contractors in the short term as we train the force. (Haas, 2015)

Implications

There is broad recognition that the present MISO force structure must grow and that it must receive effective and state-of-the-art training in web operations. Articulating the exact size of that growth and the exact nature of the training is beyond the scope of this report, but is currently under consideration by SOCOM. In light of our observations, we urge policymakers to give strong credence to SOCOM findings and provide the funding necessary to effectively enhance both manpower and capacity within the MISO forces.

5. Interagency coordination remains a fundamental challenge.

In 2011, Dr. Christopher Paul, a strategic communications expert at the RAND Corporation, testified before the House Armed Services Committee, warning that "[DoD] Subordinate formations still decry the lack of guidance for strategic communication and a lack of higher level themes and messages. . . . Room for improvement clearly remains"

(Paul, 2011). This observation came on the heels of numerous strategic communication assessments that had offered a similar conclusion and had decried the poor state of interagency coordination for "persuade, influence and inform" communications (see Paul, 2009, for a review of published strategic communication recommendations).

Based on interviews conducted for this study, five years after this testimony, the challenges remain the same. Numerous interviewees articulated a fundamental problem with receiving consistent communication themes and messages that can inform U.S. government messaging within DoD. One interviewee lamented there being "no broad national themes" for countering Russia's propaganda (interview 3, 2016). Another noted, "If I am going to influence a country, I don't know what to say that correlates to a national-level plan" (interview 7, 2016). Still another senior MISO officer recounted that shortly after 9/11, a lack of national-level themes and messages emanating from the NSC forced him to dissect presidential speeches in order to draw messaging guidance for operations in Afghanistan. He said national-level guidance had improved little in the subsequent 15 years (interview 8, 2016).

Without such guidance, different U.S. agencies, to include DoS and DoD and others, will convey conflicting themes and messages that lead to confusion and information fratricide (interview 8, 2016). One MISO officer noted that "DoD has a lot of tools, but if not orchestrated, all people hear is static." A State Department official agrees with this assessment: "Because we don't observe any discipline of messaging, there are always people saying the opposite of whatever we say it is. So, confusing messaging is part of our problem. We have met the enemy, and he is us" (interview 6, 2016).

Several interviewees complained that there is an absence of a clear lead for communications. Referring to the DIME concept (Diplomatic, Informational, Military, and Economic), one source complained: "The DIME . . . the D, M, E have very clear leads. The I has no lead. It is an orphan. The organizational construct for the U.S. government to yield information is not optimized. . . . Who is pulling it together? You tell me" (interview 7, 2016). Other interviewees noted deleterious effects that result from the fact that no single U.S. agency has the broad authority to coordinate and synchronize all U.S. government communication efforts.[8]

Implications

The study interviewees shed light on several options for improving the coordination and synchronization of U.S. government communications. Overall, the need to balance speed and consistency of messaging will require a nuanced handling of communications.

First, some interviewees recommend a Joint Interagency Task Force (JIATF) model, which can be stood up on an issue-by-issue basis. The GEC is an example of such a model. Staffed, at least lightly, by interagency offices, it has responsibility for coordination and synchronization of U.S. government communication efforts against the Islamic State. To accomplish this mission, the GEC disseminates anti-ISIS themes across the interagency offices and holds numerous interagency meetings to shape a coordinated message. In theory, as other crises emerge, the United States can authorize new JIATF structures to counter the ISIS threat. Observed one DoD officer, "You need [a synchronization hub] by functional issue rather than a single [and overarching] communications hub. The GEC focus on countering ISIS and terrorism writ large is really appropriate" (interview 7, 2016). Indeed, there is some evidence that the United States is moving in this direction. The Portman Murphy Bill, currently under consideration by Congress, seeks to authorize a new interagency organization tasked with countering Russian propaganda.[9]

Second, many individuals interviewed for this study recommended some type of regeneration of the old USIA. A former employee of the USIA and now retired State Department official recommended

---

8 As one reviewer of early drafts of this report noted, "Do we want speed or well coordinated long lead in times? How can we balance the two delicately? We recognize that this is an excellent question. Ideally, if effective coordination processes are put in place, then the interagency can promulgate key coordinating themes and messages to subordinate elements. Subordinate elements can than draw on those themes to message in real time."

9 One critique leveled against the JIATF model, however, is that there would still be a need to coordinate across JIATFs. In such a case, "Who is your synchronizer? Who would arbitrate a dispute between different JIATFs?" (interview 10, 2016).

that such a model would be critical for the United States today. He envisions an "empowered entity that does political warfare, that talks about telling America's story to the world but also aggressively goes after adversaries. . . . None of the structures, neither [Public Diplomacy] at State nor [Information Operations] at DoD nor [Broadcasting Board of Governors] do what is needed. Some have elements that would be part of an answer. This will increasingly be important" (interview 9, 2016). He also noted that such an organization would have to be "well linked to political leadership and empowered" and will need "political cover from the White House" (interview 9, 2016). Another individual observed that such an organization would need authority to disseminate messages across areas of operation or countries so that, for example, one could send a tweet to three countries without getting the approval of three ambassadors (interview 3, 2016).

The final option is to maintain the status quo, adding no new coordination structures and keeping message coordination centered at the White House. One respondent said that a USIA-type entity might be a good idea, but only if it were resourced correctly. His assumption was that such resources would never be granted, however, and so it would be "more effective if you build out multiple teams and work on coordination, rather than having all OPCON [operationally controlled] through a central point" (interview 11, 2016). Likewise, a GEC staffer opined that the current GEC interagency coordination efforts are working well, thus mitigating the need for a new agency.[10] Another GEC official seconded this observation, stating that the GEC had an interagency deconfliction issue it could not solve (interview 12, 2016). Others, however, wish to shift the control of messaging to the NSC (as opposed to DoS), but suggest this would require a significant increase in the NSC's capability. "NSC needs to be more forceful to do what the executive branch wants to do on a problem set, such as ISIS and Russia (interview 7, 2016). This individual also suggested that the United

[10] As this individual noted, "We can get around by integration and synchronization. If another agency wants to pursue a given theme or action, I can say, 'please don't do that.' And they can come back and say, no, it is important for such and such reason. If they then massage the message to adjust to my concerns, that coordination is achieved" (interview 10, 2016).

States closely examine coordination efforts in other countries. He notes that Colombia has an information secretariat, which plays a USIA-type role within the ministry of presidency, and that there would likely be value in studying this as well as other entities (interview 7, 2016).

Which of these options is best? In an ideal scenario, the NSC would effectively conduct coordination and synchronization of communications, as the NSC is the only institution that has the full authority to task subordinate departments and agencies, unless the President designates a lead agency (e.g., State).[11] Consequently, efforts to push this coordination role out to any single department, without providing full tasking authority, will provide only a limited solution. Complaints about the NSC's limited oversight of strategic communications continue even after 15 years of conflict. Short of a major overhaul of NSC operations, this portends poorly for any near-term improvements. And even if NSC did serve this coordination role, there would still be a need for a robust office that could execute strategic-level communications.[12] The JIATF model, of an issue-focused organization that

[11] It is true that the NSC can designate an agency to lead interagency coordination as is the case with the GEC. Several interlocutors within the GEC note that despite this "lead" status, the GEC "can't mandate" the actions of other agencies. However, as noted, the GEC remains satisfied that it can effectively shape the interagency via interagency meetings and senior leader engagement. It was also noted that it could work through the NSP Deputies Committee meetings were an irresolvable issue to arise (interview 12, 2016).

[12] There remains some debate as to which agency within the U.S. Government should have responsibility for strategic level influence communications. Currently, the Department of Defense has only a writ for tactical and operational level influence communications, in fitting with its Title Ten authorities. Some in the Department of Defense, however, argue that its Title Ten writ should be expanded to include broader strategic level communications. One senior MISO officer interviewed for this study argued that the military has both the resident expertise and manpower for executing influence operations. During Operations Iraqi Freedom for example, the DoD had significantly increased authorities to conduct broad and expansive communication campaigns aimed at stopping the flow of foreign fighters into Iraq. Others however believe that the DoD should remain operationally and tactically focused on supporting U.S. military campaigns and avoid broader and more strategic influence efforts. Here the argument is that moving the strategic information fight to the DoD would be an excessive expansion in DoD authorities and would militarize the information fight at a time when many key target audiences eschew military involvement. It is also argued that the U.S. Department of State can much more effectively leverage nongovernmental organizations as well as global footprint of U.S. embassies.

could be stood up as crises arise, is one potential solution. However, the process of standing up such an entity will invariably delay an effective U.S. response when it is needed most urgently.[13] That leaves the option of crafting a new agency, possibly drawing on the USIA model.[14] The benefit of such an institution would be its devoted focus to the information fight. Such an effort should be closely linked to and empowered by the White House, and it could play a key role in helping the White House coordinate and deconflict interagency communications.

References—Chapter 7

Beauchamp, Gorman, "Intellectuals Undercover," *Michigan Quarterly Review*, Vol. 40, No. 2, Spring 2001. As of November 24, 2015:
http://quod.lib.umich.edu/cgi/t/text/text-idx?cc=mqr;c=mqr;c=mqrarchive;idno=act2080.0040.216;rgn=main;view=text;xc=1;g=mqrg

Broadcasting Board of Governors, "Our Networks," n.d. As of September 19, 2016:
https://www.bbg.gov/networks/

Cooper, Helene, "U.S. Drops Snark in Favor of Emotion to Undercut Extremists," *New York Times*, July 28, 2016. As of July 30, 2016:
http://www.nytimes.com/2016/07/29/world/middleeast/isis-recruiting.html

Echevarria II, Antulio J., *Operating in the Gray Zone: An Alternative Paradigm for U.S. Military Strategy*, Carlisle Barracks, Pa.: Strategic Studies Institute, United States Army War College Press, 2016.

Garcia Jr., Colonel Victor G., "Strategic Influence: A Framework to Counter Violent Extremist Ideology," 2013, United States Army War College, Strategy Research Paper.

Goodstein, Laurie, "Muslim Leaders Wage Theological Battle, Stoking ISIS' Anger," NYTimes.com, May 8, 2016. As of July 22, 2016:
http://www.nytimes.com/2016/05/09/us/isis-threatens-muslim-preachers-who-are-waging-theological-battle-online.html

[13] Here, the CENTCOM perspective is useful, specifically that even in times of peace, it is critical to keep a standing communication hub operational so that it can effectively surge during crises.

[14] Developing such an organization would involve some significant lead time however once established, the center would in theory be situated to respond to crises at a more rapid rate.

Haas, Christopher, "Prepared Statement of MG Christopher Haas, U.S Special Operations Command Director, Directorate of Force Management and Development: Before the House Armed Services Committee, Emerging Threats And Capabilities Subcommittee," Washington, D.C.: U.S. House of Representatives, October 22, 2015. As of July 22, 2016: http://docs.house.gov/meetings/AS/AS26/20151022/104086/HHRG-114-AS26-Wstate-HaasUSAC-20151022.pdf

Hess, Pamela, "DOD to Close 'Influence' Office," UPI, February 26, 2002. As of July 30, 2016: http://www.upi.com/Business_News/Security-Industry/2002/02/26/DOD-to-close-influence-office/73281014757331/

Jacob, "That Propaganda Program Bill Clinton Praised Hillary For? It Was Considered a Failure," *Politico*, July 27, 2016. As of July 30, 2016: http://www.politico.com/magazine/story/2016/07/bill-clinton-hillary-clinton-counter-messaging-radicalization-islamists-214112

Joint Staff, Joint Publication 1-02, Department of Defense Dictionary of Military and Associated Terms, Washington, D.C., November 10, 2010.

Lewis, M. Paul, Gary F. Simons, and Charles D. Fennig, eds., "Summary by Language Size," *Ethnologue: Languages of the World*, 19th ed., Dallas, Tex.: SIL International, 2016. As of July 21, 2016: http://www.ethnologue.com

Lumpkin, Michael D., "Engaging Through Messaging: The New Global Engagement Center," *The Ambassadors REVIEW*, Washington, D.C.: Council of American Ambassadors, Spring 2016. As of July 21, 2016: https://www.americanambassadors.org/publications/ambassadors-review/spring-2016/engaging-through-messaging-the-new-global-engagement-center

Mazarr, Michael J., *Mastering the Gray Zone: Understanding a Changing Era of Conflict*, Carlisle Barracks, Pa.: Strategic Studies Institute, United States Army War College Press, 2015.

Miller, Greg, "Panel Casts Doubt Against U.S. Propaganda Efforts Against ISIS," *Washington Post*, December 2, 2015. As of July 30, 2016: https://www.washingtonpost.com/world/national-security/panel-casts-doubt-on-us-propaganda-efforts-against-isis/2015/12/02/ab7f9a14-9851-11e5-94f0-9eeaff906ef3_story.html

Moore, Charles, "Testimony Before the House Subcommittee on Emerging Threats and Capabilities: Witness Statement of Brig Gen Charles Moore, Deputy Director for Global Operations, Joint Staff," Washington, D.C.: U.S. House of Representatives, October 22, 2015. As of July 22, 2016: http://docs.house.gov/meetings/AS/AS26/20151022/104086/HHRG-114-AS26-Wstate-MooreUSAFC-20151022.pdf

Obama, Barack, *Developing an Integrated Global Engagement Center to Support Government-Wide Counterterrorism Communications Activities Directed Abroad and Revoking Executive Order 13584*, Washington, D.C.: The White House, Office of the Press Secretary, Executive Order 13721, March 14, 2016.

Parish, Karen, "CENTCOM Counters ISIL Propaganda," *DoD News, Defense Media Activity*, July 6, 2016. As of July 22, 2016:
http://www.defense.gov/News-Article-View/Article/827761/centcom-counters-isil-propaganda

Paul, Christopher, *Getting Better at Strategic Communication*, Santa Monica, Calif.: RAND Corporation, CT-366, 2011. As of July 22, 2016:
http://www.rand.org/pubs/testimonies/CT366.html

———, " 'Strategic Communication' Is Vague: Say What You Mean," *Joint Force Quarterly*, No. 56, First Quarter 2010.

———, *Whither Strategic Communication? A Survey of Current Proposals and Recommendations*, Santa Monica, Calif.: RAND Corporation, OP-250-RC, 2009. As of July 22, 2016:
http://www.rand.org/pubs/occasional_papers/OP250.html

Seib, Philip, "CENTCOM's Digital Engagement Tries to Counter Extremists," *USC Center on Public Diplomacy, CPD Blog*, Los Angeles, Calif.: University of Southern California, August 26, 2010. As of July 5, 2016:
http://uscpublicdiplomacy.org/blog/centcoms_digital_engagement_tries_to_counter_extremists

Silverman, Jacob, "That Propaganda Program Bill Clinton Was Praised For? It Was Considered a Failure," *Politico*, July 27, 2016. As of February 3, 2017:
http://www.politico.com/magazine/story/2016/07/bill-clinton-hillary-clinton-counter-messaging-radicalization-islamists-214112

Suebsaeng, Asawin, "The State Department Is Actively Trolling Terrorists on Twitter," *Mother Jones*, February 27, 2014. As of February 3, 2017:
http://www.motherjones.com/politics/2014/02/state-department-cscc-troll-terrorists-twitter-think-again-turn-away

Stengel, Richard, Testimony Before the 114th Congress, U.S. House Committee on Foreign Affairs, July 13, 2016. As of July 22, 2016,
http://docs.house.gov/meetings/FA/FA00/20160713/105223/HHRG-114-FA00-Wstate-StengelR-20160713.pdf

United States Department of State, "Launch of the Sawab Center," Washington, D.C.: Office of the Spokesperson, July 8, 2015. As of July 22, 2016:
http://www.state.gov/r/pa/prs/ps/2015/07/244709.htm

United States Information Agency Overview, 1998. As of July 22, 2016:
http://dosfan.lib.uic.edu/usia/usiahome/oldoview.htm

White House, Executive Order 13584, Developing an Integrated Strategic Counterterrorism Communications Initiative, September 9, 2011.

White House, Executive Order 13721, Developing an Integrated Global Engagement Center to Support Government-wide Counterterrorism Communications Activities Directed Abroad and Revoking Executive Order 13584, March 14, 2016.

CHAPTER EIGHT

Effective Statecraft and Integration of Measures Short of War

This chapter explores the strategic requirements for effective political warfare, the organizational and operational requirements for effective responses by the U.S. government, and the requirements for coordinated action with other partners—including cases in which a coalition, alliance, or other sovereign country plays the leading role. The chapter concludes with recommendations for the U.S. military, and U.S. special operations forces (SOF) in particular as they provide many of the capabilities used in political warfare.

The research for this chapter was conducted primarily through 40 interviews with currently serving and retired officials from the U.S. State Department, other U.S. policymaking entities (such as the National Security Council and the Department of Defense), and U.S. allies and partners, including European, Canadian, and Australian civilian and military officials. The interview research was supplemented by reviews of relevant official documents and scholarly literature.

Strategic Requirements for Effective Offense and Defense in Political Warfare

To respond effectively to aggression that falls short of conventional war, the U.S. government needs not only a suite of methods that achieves the desired effects, but an overarching strategy and set of organizing principles for applying those methods. In most cases, the aggression may be directed against U.S. partners, allies, or vital interests abroad rather than the U.S. homeland. This introduces a further complexity in the form of a requirement to coordinate the U.S. strategy and approach with those partners and allies.

The United States, as all countries, must have some organizing principles to determine its national interests, their relative values, and the lengths to which it will go to defend those interests. Governments must also decide when and where not to intervene, as not every challenge short of war requires a response. Former U.S. Secretary of State Henry Kissinger, in his recent book *World Order*, argued that a first step in establishing a strategic framework is to determine the hierarchy of U.S. interests, the goals it intends to pursue, and the values it seeks to defend. Otherwise the U.S. government cannot hope to craft an effective approach to the contests that roil the international system today (Kissinger, 2014, pp. 372–73).[1] This framework is particularly important to serve as a compass to direct the U.S. government's attention to those areas where creeping, ambiguous attempts at subverting the established order or challenging U.S. or allied interests should be countered—ideally early to prevent or preempt a worsening crisis and also ideally by the least costly effective means available.

The multiplicity of threats at any given time, and certainly in today's turbulent world, requires policymakers to scale U.S. responses appropriately in relation to both the threat to U.S. interests and also its available resources. Resource constraints are another factor forcing clarity of purpose and efficacy in approach. The issue of resource constraints has become more salient in recent years. As MIT political scientist Barry Posen argues in his critique of "Liberal Hegemony," the tendency to rely on military levers of power will exhaust the United States (Posen, 2014). Yet failing to deter and halt aggressors who have

---

[1] Each U.S. administration does endeavor to answer these questions in the form of a national security strategy, to greater or lesser effect. Typically, though, the resulting strategy documents are broad statements of objectives that do not closely tie them to the ways and means of achieving them, and rarely acknowledge the implicit choices being made due to limited means or interests. The Quadrennial Defense Review is a congressionally mandated assessment that is limited to the military element of national power and aims primarily to establish defense program priorities. In the past decade, the Quadrennial Diplomacy and Development Review has sought to do the same for diplomacy and development programs.

the ability to destabilize critical allies or entire regions is a risky and ultimately costly gambit as well. Therefore, the task of the policymaker is to define U.S. interests clearly and forge a strategy to defend them.

The salient features of an effective offense are (1) an enunciated strategy based on shared values to garner international approval and support; (2) cost effective tactics, which entails both low-end military tools and less-expensive diplomatic, informational and economic measures; and (3) a burden-sharing scheme to distribute tasks to those with the greatest competence, insight and efficiency in particular activities. The primary features of an effective defense follow the same principles above along with (1) proactive use of detection and preventive measures to deter would-be aggressors, even if this protective umbrella carries some risk of overstretch; and (2) heavy investment in remedying economic and political vulnerabilities to bolster a target country's identified vulnerabilities.

A retired senior U.S. diplomat interviewed for this study argued that the strategy must be clearly broadcast as an affirmative and inclusive call for promoting and defending a rules-based international order. This positive formulation does not frame the overarching U.S. goal as conducting "warfare," but envisions robust proactive measures as needed, including coercive ones, as well as deterrence and defense. This strategy should include a disposition to undertake necessary reforms. Articulating such a vision would provide impetus for a large coalition of partners willing to undertake various roles in a coordinated effort.

Currently, the post–World War II international institutions and the basic Westphalian rules by which sovereign states manage competition and seek to resolve their disputes is under strain from both domestic and foreign pressures. Thus, promotion and defense of the order may also entail reforms to enhance its essential stabilizing functions, which rest on its defenders' consent that the process possesses a basic legitimacy. As Kissinger notes in his recent book, the international system is under strain because the dynamics of global capitalism are de-linked from the dynamics of state control, and challenges to the old order arise from both domestic frictions and revisionist state powers such as Russia, an emerging power China, and alternative models posed by radical Islam.

This study has illustrated how adversaries seek to use low-cost measures short of conventional war to achieve their aims. The United States should adopt the same approach of seeking less costly rather than more costly means of response, provided they are effective. Recent works by David Gompert and Robert Blackwill—as well as the seminal *Soft Power* by Joseph Nye—have explored nonmilitary tools (including reputational credibility and attractive values) and how they might be used in these circumstances. Rather than serve as the global 911 service, the Nixon Doctrine's relied on a network of strong regional partners to police given areas (Brands, 2014). The Reagan Doctrine's use of nonstate "freedom fighters" to contest Soviet aggression and proxy states represents another type of cost-saving alternative to large-scale U.S. military presence or interventions, yet they also avoid the high risks of quasi-isolationism that some offshore balancing strategies might entail.

To summarize, the United States will be in a stronger position if it can determine what it stands for, and not just what it is against. Such a values-based approach may garner more internal and external support from those who are motivated by its potential benefits and may thereby enhance the chances of its success, as numerous interviewees argued. Second, therefore, given the multiplicity of challenges, the United States must exercise discipline to ensure that it can in fact defend and promote those interests it deems vital and critical. Third, the lowest-cost effective approaches should be given priority over more expensive ones, and these can include early preventive and counter-escalatory approaches if the United States can develop sufficiently successful formulas to produce the desired return on investment. Finally, burden-sharing among partners yields a more effective and efficient approach.

Organizational Requirements and the Leading Role of State

The need to devise effective approaches to political warfare challenges received wide support among the dozens of civilian and military U.S.

and non-U.S. officials and experts interviewed for this study.[2] Among the challenges, Russia and ISIL were cited as two examples of contemporary actors whose transgressive behavior and transregional reach are considered sufficiently threatening to require collective and multifaceted responses. Yet, these practitioners and experts note, the U.S. government chronically struggles to create an effective "whole-of-government" response that employs a full panoply of the tools at its disposal, not to mention leverages the resources of the wider society. As one interviewee noted, "An ever-present problem with 'gray zone' activities is that no U.S. government agency has the lead, and we have been reluctant to develop a forcing function—an entity with the authority to ensure the actions and activities of our various governmental agencies are integrated and supportive of presidential decisions" (interview with U.S. government official in Tampa, April 19, 2016). That, in fact, should be the organizational goal for countering political warfare.

While wide agreement exists that an effective response is needed to address threats short of conventional war, interviewees frequently cited a lack of day-to-day, routine coordination among the various U.S. departments and agencies. No military-type command and control is exercised over the entire government. The growth of the National Security Council staff in recent years may be attributed in part to the effort to replicate and extend this degree of control, but this development has been criticized as leading to overinvolvement in the tactical levels of policy execution while still failing to produce effective overarching strategies for policy guidance. That is, this development has still failed to link policy objectives to clear, feasible, and resourced plans to accomplish the desired aims (Ries, 2016). Thus, the growth of the NSC staff to a historical high of 500 personnel replicates many functions

---

2 A number of analysts have used political warfare to describe adversaries' activities. Center for Strategic and International Affairs Fellow Anthony Cordesman recently used the term to describe Russian behavior in Syria (Cordesman, 2015). Similarly, Asia experts—including Princeton professor Aaron Friedberg and others—use the term to refer to China's campaign to wield political influence abroad against the United States and its allies (Parameswaran, 2015). Some analysts do argue that therefore political warfare should also be employed as part of the American strategy. For example, Council on Foreign Relations Senior Fellow Max Boot and Brookings Institution Fellow Michael Doran called for the United States to rebuild its own capacity to wage political warfare (Boot and Doran, 2013).

performed elsewhere in government without being sufficiently large or broad enough in scope to develop detailed, integrated strategies for each critical problem.

Designating State as the Lead

The offensive and defensive approach to political warfare outlined above is tantamount to old-fashioned muscular statecraft, in which political knowledge of countries and regions is married to diplomatic skill, backed as needed by the threat of force or coercion. The leading role for designing and implementing such an approach would naturally fall to the State Department as the Cabinet traditionally considered as first among equals.

When asked which department or agency should lead the overall response to conflicts short of conventional war, the majority of those consulted during this study, both civilian and military, believed that the State Department is the appropriate entity given that, as one interviewee stated, "it is the one part of the government created expressly to do that, to represent and advance the nation's foreign policy goals overseas" (Ries, 2016). A military interviewee said, "DoS should have the lead in executing the strategy while being supported by the remainder of the government" (interview with U.S. military official, April 18, 2016). The reason given is that the activities documented in the earlier chapters show that political warfare or "gray zone" challenges are heavily nonmilitary in nature, and the primary responses are also likely to be nonmilitary or at least nonkinetic in nature.[3]

As part of its duties, the State Department tracks the political, economic and cultural developments of countries around the world, and engages in a wide array of interactions to promote U.S. interests and goals. The President, supported by his National Security Council, will shape the policy guidance, but the Secretary of State is historically

---

[3] Even the military contributions are likely to be nonkinetic, ranging from signaling and deterrent activities such as forward deployed presence, patrols, exercises, and military assistance to friends and allies, to more active but low-visibility, clandestine, or covert measures undertaken normally by special operations forces or intelligence services with heavy reliance on local agents. Early and ongoing intelligence collection and analysis is required to detect and interpret the adversary's typically ambiguous and/or concealed activities.

the acknowledged lead cabinet member in matters of foreign affairs. The guidance issues from the White House, and the State Department crafts specific policies and programs to carry it out. Such guidance from the White House includes communications guidance, as noted in the previous chapter. But the machinery to develop and execute detailed responses to nonmilitary threats does not exist at the White House, and by portfolio and mandate would properly reside at State.

In addition to this overarching role in calibrating and coordinating the application of DIME measures as a coherent response, due to its knowledge of the countries and regions, the State Department, along with the USAID, possess the country familiarity and niche specialties to diagnose and address the political aspects of the conflict. Most political warfare threats gain traction from conditions inside a country or region, and measures to address them must necessarily be tailored to individual countries' characteristics. As one State Department official stated, "DoS should work on conditions that give rise to political warfare, to address the common variables that we see in conflict areas, whether prone to ISIL or other FTOs [foreign terrorist organizations] or Russia related or other state-generated conflicts" (interview with U.S. State Department official, Washington, D.C., April 12, 2016). Another State Department official emphasized that the coordinating role must be a collaborative one, given the important roles of defense and intelligence: "State should lead but collaborate with DoD and the intelligence community in both the Washington and chief-of-mission environment. . . . The best mechanism on the military side will be SOF capabilities working together, with embassy oversight pulling things together" (interview with U.S. State Department official, Washington, D.C., April 12, 2016).

Most of those interviewed judged that the U.S. country team, led by the U.S. chief of mission, was, by itself, an inadequate interagency organizing construct. Why? The most significant foreign policy challenges span more than one country, and to be effective, an overarching approach must be developed, with appropriate allowances for variation.

That said, the one standing interagency construct at the State Department is the country team, and when led by a capable chief of mission, it routinely achieves the type of whole-of-government synchronization that is particularly required to confront threats that are also using a full DIME set of tactics and resources. The presidential authority conferred on the chief of mission does create the needed "forcing function," enabling the ambassador to harmonize the array of activities that are overseen by the country team's various departmental and agency representatives for best effect. The country team is also a vital source of knowledge and thus must be integral to the effort to craft a regional approach. The country team's diverse members can also, by their posting in the country, gain a deep and collective appreciation of the challenges and environment to craft the best possible approach to addressing the problems in concert with the sovereign host government. However, one veteran U.S. diplomat noted a worrying tendency for U.S. departments and agencies to still send personnel around the world to conduct missions that are not coordinated with the country team (interview with U.S. State Department official, Washington, D.C., July 6, 2016).

Despite the many strong attributes of the country-team construct, therefore, the majority of interviewees agreed that two related functions must be performed above the level of the country team, and that both can be performed by an entity led by State: (1) interagency policy planning and coordination, and (2) coordination for regional problems that exceed the reach of one country team.

Interviewees differed regarding whether additional written authorization is needed to ensure or codify State's leading role in this realm. One currently serving diplomat believes that no additional organizational changes are required, and no additional government directives are needed, given the long-standing primacy of the State Department in conducting the nation's foreign policy. The only requirement, in this view, is that the White House enforces this primacy, by itself operating through the State Department and ensuring that others do so. Others believe more explicit guidance is needed, given the nearly constant communications and travel by staffers of all the U.S. government's departments and agencies, and the tendency of the very large military bureaucracy to fill gaps that it perceives.

Other respondents agreed that explicit direction from the White House for State leadership on specific issues is necessary. A military interviewee agrees that it is desirable that "DoS be afforded the requisite authority to integrate and synchronize the variety of elements of national power in support of an approved "gray zone" campaign (interview with U.S. military official, April 18, 2016). A senior retired diplomat with experience in multi-country interagency policy execution believes a presidential decision directive is required to explicitly enforce a State Department-led process and create structures as necessary to plan and implement whole-of-government strategies. "The president needs to declare the national policy on a given issue so that the chiefs of mission know where they fit, and everyone at the interagency table knows. It is the equivalent of the president's [chief of mission] letter. Then, State can coordinate planning and execution" (interview with retired senior U.S. diplomat, Washington, D.C., June 20, 2016).

State Department–led interagency coordination has occurred in the past through various organizational structures. One interviewee recommended that the Under Secretary of State for Political Affairs should be the "belly button" for coordination of U.S. government policy response to Russian and Chinese revanchist activities. He cited the precedent of former Under Secretary of State for Political Affairs Tom Pickering as the first of three coordinators of U.S. government support to Plan Colombia, a decade-long multifaceted program to assist the Colombian government's battle against drug trafficking and guerrillas that spanned military and counternarcotics assistance from both State and DoD; judicial, law enforcement, and intelligence support; development aid; and an Andean free trade pact. With congressional backing, the plan also promoted democracy, pluralism, and human rights. Another more recent example is the State Department's Special Representative for Afghanistan and Pakistan, who, along with a robust team, has guided interagency policy planning and execution for the past decade. (However, that office has at times encountered difficulties, some of them personality-based, in working with the White House and the Pentagon, which have also had czars and internal structures simultaneously devoted to this effort.)

Overall, these previous State-led structures implemented the broad policy guidance issued by the White House and formulated through the principals, deputies, and interagency policy committee process. This organizational structure has been an occasional rather than routine occurrence, and many of those interviewed believe it must become a standing practice. The other rationale for a State Department–led structure is that it can articulate the broader U.S. national interest behind a concerted approach, one that transcends narrower military, economic, or other rationales, in its role representing U.S. foreign policy. "We have allowed the TTP [the Trans-Pacific Partnership] to become about car parts, when it is about a strategic values-based response. We have to start talking about the attack on the international system," an interviewee stated. "Putin is pursuing measures short of war very successfully to tear apart Europe. Taken in totality, these threats to the international order definitely amount to an existential threat" (interview with retired senior U.S. diplomat, Washington, D.C., June 20, 2016).

Operational Requirements to Fill Seams and Gaps in the U.S. Government

If a State Department–led organizational structure is deemed the most appropriate entity to orchestrate an interagency response to the political warfare or gray zone challenges, the question is whether the requisite capabilities and practices are in place to effectively implement this approach as a routine matter. Due to its vastly greater manpower and budget, DoD very often fills the gaps due to the lack of capacity or capability of the State Department. As one interviewee noted: "Oftentimes, DoD has been the default U.S. government response—partly because it has the most capability and resources. [Yet] to be successful, our government's response to the challenges will require an unprecedented level of interagency coordination with all elements of national power synchronized." Thus, he concluded, in addition to evolving the organizational model for "gray zone" responses, "The U.S. also needs to grow non-military capabilities—current gray zone actions are overly militarized" (interview with U.S. government official, Tampa, Fla. April 19, 2016).

This military official finds ready agreement among many at the State Department, who believe that this default to relying on military organizations to lead results in the prioritization of military approaches and military-oriented thinking. A senior diplomat asserted that the outcomes in Iraq and Afghanistan to date reflect this over-emphasis on military measures at the expense of critically needed nonmilitary approaches to achieving the endgame of relative peace and stability and sufficient capacity for those governments to tend to their people's needs.

These are not new critiques, but in the view of both military and civilian officials, the continuing failure to address them systematically will preclude successful responses to the hybrid, asymmetric, irregular, and political challenges of today's environment. The 6,500-person diplomatic corps does not need to become as large as the 1.3 million active-duty armed forces, but it needs sufficient personnel to accomplish key tasks, conduct coordinated planning, undergo training, and receive at least some of the educational opportunities that are routinely afforded to the nation's uniformed services. Military personnel receive the opportunity to earn advanced degrees during their careers, and they have up to a year to prepare for deployed assignments, normally visiting their predecessor units one to three times during that year. In contrast, civilian personnel receive a few weeks training and are lucky if they arrive in their post before their predecessor departs.

Despite some additional hiring in the Foreign Service ranks over the past decade, Congress has routinely declined to approve multiple administrations' requests for more robust funding, even though in 2016 the State Department's budget was $27 billion versus $516 billion for DoD, of a total $3.5 trillion in federal government spending (Inside.gov, 2016).[4] A recent survey of congressional staff indicated that there is little sense of clarity about the perceived gaps at State or sense of urgency to remedy them. "A majority of respondents (61 percent) said they and their offices were not interested in the issue of professional formation and development of Foreign Service members. . . .

---

[4] Despite this overall disparity in funding, the budget for civilian public diplomacy and information, to include the seven civilian-run international radio and television stations overseen by an independent federal agency, the Broadcasting Board of Governors (BBG), is larger than the MISO budget. The BBG FY15 budget was $742 million, with a staff of 3,590 (Broadcasting Board of Governors, n.d.).

At the same time, [only] 33 percent said they thought there is sufficient training" (Kralev, 2013).

The other issue, somewhat related to resources, is whether the State Department's current practices are conducive to performing the roles described here. These gaps in practices are illustrated by two examples, one of which did not result in the needed coordination and another of which holds promise of doing so.

In the first example, a military command sought State Department participation in a multi-week planning session to develop joint planning guidance. The relevant bureau sent one senior official, but only for one day. The command's leadership visited Washington and briefed State Department officials three times, but, according to a senior military adviser assigned to the State Department, the diplomats found the 50-page briefing deck dense and difficult to understand. The command subsequently produced a white paper, at the adviser's suggestion, and then an executive summary. Ultimately, the Pentagon decided that the matter was sufficiently urgent to warrant producing a DoD plan. The military officials involved assured State Department officials that ample consultation would be conducted before the plan's implementation. Yet the end result was that no overarching political-military plan was drawn up, and several of the interviewees indicated that there is disagreement over some of the proposed activities. One embassy, approached with a request to approve an activity within three days, balked at the short timeline as well as the proposed activity.

One of the military officials involved expressed frustration at the lack of State Department involvement in crafting the plan. "Instead of criticizing our effort, why not participate? We are happy to follow, if you will lead," said a U.S. military official who was involved with the extensive planning effort. At one level, this example illustrates the lack of sufficient State Department personnel to participate in weeks-long planning efforts, yet it may also indicate a lack of commitment to interagency planning. The State Department does assign a foreign policy advisor to major military commands, but that person typically engages with the command group and commander to provide senior-level advice and does not have the time to participate in the military's lengthy planning cycles. The result of such lack of coordination, however, according to a senior diplomat, is that "the theater campaign plan contains very little that is of use to or relevant to the chief of mission" (interview with U.S. official, Washington, D.C., July 6, 2016).

The second example illustrates a more constructive process, which could point toward a more integrated government effort. A Russian Information Group has been created, led by the State Department, with weekly videoconferences involving all relevant government departments and agencies, including those in the field. This working group identifies and discusses issues and exchanges information. A monthly meeting convenes the department directors and the military O-6 level to determine what issues they can address and which items should be elevated to the deputy assistant or assistant secretary and commanding general at quarterly meetings. This process has just begun, but participants are hopeful that it will be a productive and ongoing forum.

A similar effort to start a government-wide public diplomacy working group did not gain traction for lack of follow-through at senior levels. The commitment of senior leaders to such processes is essential if they are to function as intended. An interagency forum does not guarantee that a consensus or compromise will be found, of course. For example, several diplomats expressed wariness about U.S. military activities in the information realm, particularly when the topics concerned the delicate internal affairs of a country, such as minority populations.

One State Department official suggested that the most useful and palatable element of the newly redesigned MISO (Psyop) brigades is the Regional MISO Team, which conducts research and analysis to attribute adversary disinformation, which the host military or government then publicizes or uses as best serves its aims. A Baltic Working Group has also been formed to draw together the various DoS and DoD elements with programs in this subregion.

An interagency plan and approach to a given region will require both an overarching concept and lines of effort as well as the necessary variances that correspond to particular countries' unique characteristics. For example, even in the case of the Baltic countries, a one-size-fits-all approach is not appropriate. Both Estonia and Latvia have large Russian minority populations, while Lithuania does not. Country teams and the countries themselves are invaluable sources of information about these particularities. A number of interviewees note that State, Defense, and the intelligence community should work together to develop a common understanding of threats in each region, subregion, and country, as well as the most effective responses. That common picture of likely scenarios would provide the necessary baseline for determining needed actions. The above ad hoc organizational structures may represent partial attempts to establish such a baseline, but there has been no explicit guidance issued to develop a whole-of-government campaign plan even in the case of Russian aggression in Europe. Without that kind of guidance, collective action toward a common objective will be difficult to achieve.

Interviewees made a number of recommendations that would, in their view, enable the State Department to routinely play a more effective leading role in coordinating interagency whole-of-government efforts:[5]

1. Strengthen the planning culture at the State Department by setting up dedicated planning cells with planning as their primary responsibility. The policy planning staff, led in the Cold War era by George Kennan, was historically the locus for overarching plans, but numerous interviewees suggested that the regional bureaus need dedicated planning cells for region-specific planning that align resources and programs to achieve needed operational outcomes. State's Bureau for Conflict and Stabilization (CSO) has developed numerous planning tools, such as the Interagency Conflict Assessment Framework, but it is currently 25-percent understaffed, according to a bureau official. The CSO could become a planning center explicitly dedicated to supporting the planning needs of the regional bureaus and embassies. Its personnel could be seconded to the regional bureaus and country teams that require detailed political warfare analysis and plans.
2. Require in-depth training in interagency processes for diplomats to enable them to effectively lead interagency groups in both

[5] These recommendations were developed from interviews with current and former State Department personnel.

planning and implementation. One diplomat recommended a mandatory six-month course at the National Defense University for all newly minted Foreign Service officers for an immersive interagency education. An alternative would be to create an interagency program at the Foreign Service Institute, but NDU might be rebranded as an interagency organization as it currently hosts a very high number of both interagency and international attendees. As a parallel initiative, he recommended that State break down the rigid "cone" structure whereby Foreign Service officers are shunted into political, economic, public diplomacy, consular, or management stovepipes, which reduces the opportunities for many, particularly those outside the political cone, to become multifaceted senior leaders.

3. Elevate and incentivize the position of foreign policy advisers, or POLADs, who serve assignments at military commands around the world. This has, in fact, been happening gradually in the past decade. Recently, 400 Foreign Service officers applied for the approximately 40 positions available (interview with U.S. official at the State Department, June 24, 2016). If the State Department routinely selects its best officers for these jobs, and then promotes them to higher positions, including ambassadorships, then the position will become increasingly valued as a means to connect the U.S. military commands not only to State but also to the wider interagency community. Recent military commanders have elevated their POLADs to a position co-equal to that of the military deputy commander, including at European Command (EUCOM), and they have held the number two positions at Africa Command (AFRICOM) and Southern Command (SOUTHCOM). CENTCOM has relied on senior Foreign Service officers not only for expertise in how Washington works, but also for country-specific and functional expertise outside the military lane.
4. Establish a close and continuing relationship between the assistant secretary of state for each region and the U.S. geographic combatant command (COCOM) that oversees military activities in that region. The regional bureaus are functionally the

equivalent to the military commands, and one retired senior diplomat believes that they need to behave as such and be treated as such. The under secretary of state for political affairs, who oversees all the regional bureaus, should also be regularly engaged with the four-star COCOMs, particularly when the issues involve transregional threats that cross COCOM boundaries and require more than a purely military approach. Any military approach should ideally be nested under an overarching State Department–led U.S. government plan. Given the military commands' vastly greater manpower and training in planning, this is currently out of reach for the State Department. But without it, the challenges to threats short of conventional war will go wanting for a coordinated interagency approach.

5. If the State Department is to take a leading role in crafting and coordinating responses to political warfare, its leadership positions must be occupied by seasoned, qualified diplomats with deep knowledge of the countries and regions in question. Thus, it would be imperative to reduce the ever-expanding number of political appointees that have been assigned to the department's most senior, critical, and sensitive positions. Career Foreign Service officers have frequently made this recommendation, to little effect. Usually, ambassadorships are awarded to campaign donors, but in the best-case scenarios, these are in countries requiring little day-to-day or strategic crisis management—or else the bilateral relationship is managed largely in Washington. The most worrying trend, pointed out by several interviewees, is the growing practice of placing political appointees into the most senior positions of assistant secretary and above. According to a report by the American Academy of Diplomacy, career Foreign Service officers hold only 25 to 30 percent of those positions today versus upwards of 60 percent in 1975.[6] This has a ripple effect, depriving the department of its most senior talent at

[6] This is the first issue singled out by the American Academy of Diplomacy's report, "American Diplomacy at Risk," April 2015, p.15.

the top, demoralizing the most talented officers, and—because the appointees have a role in the appointment of ambassadors—diluting the ability of those who know their peers to promote the best among them to the critical chiefs of mission positions.

6. Ensure that the regional and functional bureaus within the State Department operate synergistically to confront threats short of war. As the senior under secretary, the Under Secretary of Political Affairs is the logical person to ensure this synergy. While the department's regional bureaus are the logical places for planning and coordination of integrated approaches to regions or subregions, the department's functional bureaus contain many of the tools for increasing countries' resilience and confronting threats short of war. These bureaus include Conflict and Stabilization Operations; Democracy, Human Rights and Labor; Population, Refugees and Migration; International Narcotics and Law Enforcement (which includes police training programs); and Political-Military (which includes foreign assistance and training of peacekeepers). Diplomats acknowledge that these separate programs and experts need to be more effectively pulled together into a coordinated effort that is tailored for each country and region (interview with U.S. State Department official, April 25, 2016). A departmental directive may be needed to ensure that regional bureaus and the under secretary for political affairs can achieve—and if needed, compel—the desired coordination with the functional bureaus. This may require revoking the ability of functional bureaus to nonconcur. Programs beyond State, such as the National Endowment for Democracy, form a vital part of democracy promotion efforts. A more operational culture could be fostered by instituting mechanisms to chart progress and identify obstacles. The integrated regional strategy for each region should ideally reflect the contributions of all civilian capabilities.

Partnership Requirements for Coordinated Action with Other Countries

Instituting a more holistic and collaborative U.S. government effort is but one aspect of an integrated approach, for the U.S. government does not act in a vacuum. In all but a very few circumstances, the responses that the United States would mount to any political warfare challenges would be carried out in support of others, and possibly alongside allies or partners.

Thus, detailed planning necessarily requires engagement with the government of the country that is directly threatened. Once a common plan is formulated, assistance can be delivered bilaterally or as part of a coalition or alliance. The range of military activities in which others may seek assistance can range widely, from making deterrence more credible; to increasing military presence, posture, exercises, or aid; to activating defense plans and sending signals regarding measures to be taken in the event of military aggression. Subversive or low-grade activities can be actively countered through covert, clandestine, or overt means.

Countering aggression can also take positive, reinforcing forms, and these are generally referred to as strengthening a partner's resilience. For example, measures to reduce vulnerability to subversion or coercion can include redressing the economic, social, or political grievances of populations or minorities that might be targets of recruitment by an adversary. The general cohesion of a society is also a factor in resilience: Any fractured society can be further rent by outside powers that play on, distort, or exacerbate these cleavages. In Ukraine, for example, Russians portrayed government supporters as fascists or Nazis; while in Europe, both far-right and far-left political parties exploit anti-immigrant, anti-European Union, and anti-NATO sentiments. Many threatened countries, such as Estonia, formally embrace and plan for resilience as a means of national defense. In all of these cases, support for building national resilience is another type of external support that friends and allies might provide.

The following sections use Estonia, NATO and Britain as examples of targeted countries, coalitions and individual allies in order to provide detailed discussions of the respective roles they may play in an integrated approach to countering political warfare. While no example serves as a comprehensive illustration, a number of these practices and approaches could apply to countering subversion and coercion exercised in the Middle East or Asia by Iran, China, or other actors.

Estonia as an Exemplar of Host-Nation Considerations

Estonia is an example of a country that has been subjected to a steady campaign of threats and pressure from Russia, as the earlier case study described. In addition to seeking robust support from NATO to serve as a deterrent and to signal that the alliance will act on its Article 5 commitment to defend its member country in the event of a Russian military intervention, Estonia is undertaking its own whole-of-government effort to prepare for a range of contingencies. Estonia's leaders believe that the Russians could seek to provoke or stimulate unrest in its Russian-dominated northeastern region and use that pretext to intervene to protect fellow Russians in the guise of a "peacekeeping" force. An ambiguous stand-off could pose a serious test for NATO, if the provocation were staged clandestinely by plainclothes Russians and attribution could not readily be established. Even if NATO is much more alert to such a scenario after the clandestine activity in Crimea, the current pressures facing Europe could prompt an opportunistic Russian leader like Putin to test the alliance, some Estonians fear. They point to just such an attempt by the Soviet Union in 1924, when the Communist International (COMINTERN) attempted to carry out a coup in Estonia by infiltrating Soviet GRU agents in hopes of sparking a wider uprising with the help of the Estonian Communist Party.

An Estonian national security official related various scenarios that the government is preparing for, which range from a military intervention that Russia could launch if it feels threatened elsewhere to "continuing to foment latent conflict that it can heat up if necessary through subversion, economic pressure, IO and cyber." While Estonia is a small, flat country of 1.5 million people, it seeks to harden itself as much as possible. It is one of five NATO members that devote the 2-percent minimum of its GDP to defense spending. Moreover, its National Security and Defense Coordination directorate has conducted government-wide exercises involving finance, education, justice, culture, and other ministries, as well as the country's president, prime minister, and members of parliament. "Our conception is that it must be a whole of government and whole of nation effort," the senior Estonian official said. "Our view is that in different stages of the conflict or different types of conflict, the military can be the supporting effort." He described the strategic guidance and legal framework to "protect against the possibility that the government will become rudderless. This is to ensure that they cannot win by decapitating the government" (interview with senior Estonian official, Tallinn, June 3, 2016).

Sitting right on the border of Russia, Estonians are literally eyeball to eyeball with Russia's 76th Regiment and intelligence services. The two countries' border guards do cooperate in laying out buoys each spring to mark the border that runs 98 kilometers through waterways and flat land. A tour of the Estonian border stations reveals a network of high-tech border posts, where agents watch fishermen ply Lake Peipus and the Narva River while monitoring the dry riverbed that runs right into Narva city. A walking bridge connects workers, tourists, and shoppers, as does a multi-lane bridge into Ivangorod, Russia. This northeast area includes mines, factories, and the main electricity generating plant for the country—all potentially lucrative military targets. Next to the power plant, a tumbledown neighborhood of weekend cottages, nicknamed Venezia, is honeycombed with canals leading into the river and straight across to Russia.[7]

These physical characteristics illustrate how easily the northeastern border can be penetrated either overtly or in clandestine fashion. The Narva population, according to Estonian officials, is not actively pro-Russian but has suffered disproportionate unemployment and the dislocating effects of migrating from the top of the social strata to the bottom. "Mostly they live in their own [Russian-speaking] information space and just want to be left alone," a U.S. special operator said. An Estonian official concurred: "They wear the St. George ribbons and show the Russian flag, but it is more a sign of their separateness."

---

[7] These observations are drawn from a tour of Narva and the Estonian-Russian border area, June 2, 2016.

The Estonian government has taken some measures to reach out to its Russian population. The Estonian public television network recently launched TVE+, a Russian-language channel, but its programming is far less slick than the Russian offerings. Similarly, the government has hosted job fairs in Narva, but the positions rarely match the salaries of the higher-paid factory work. Additional measures to assimilate the Russian population might increase Estonia's resilience to Russian political warfare, but Estonians view their national identity as part of their concept of national defense. Estonian citizenship requires that Estonians speak the national language, and secondary schools must teach in Estonian.

The whole-of-government effort extends to the Estonian military, which includes a special operations unit that deployed in combat missions to Afghanistan to gain experience and political capital for the potential day it might need to call in the favor from the United States. Estonia's Special Operations Force works closely with the Estonian Defense League (EDL), a volunteer civilian defense organization that was revived when Estonia regained its independence. The EDL reports to the Estonian Ministry of Defense, which serves as a conduit for U.S. assistance. A former head of the EDL echoed this societal model of national defense, flipping the U.S. paradigm on its head: "Civilian resistance is necessary for the military to survive. Civil society is critical for deterrence and defense at all stages from Phase 0 through occupation. Civil defense forces form a critical bridge between society and the military. If this bridge works, civil resistance can succeed in an emergency. Civil society will form the local authorities, even as the elected leadership is killed" (interview with Estonian Ministry of Defense adviser, Tallinn, June 3, 2016).

Estonian SOF and the security services work closely with law enforcement and the border guard in peacetime, and both the EDL and border guard are forming additional units. The border guard cannot hope to weed out all Russian agents from the flood of 8,000 to 12,000 civilians coming and going each day, any more than the tiny Estonian military can hope to hold off Russian tanks until NATO forces arrive—especially given the airspace denial that Russia has achieved by moving its S-400 anti-aircraft system into Kaliningrad on the Baltic Sea. In the worst-case scenario, if its own hardening efforts and NATO's umbrella prove an insufficient deterrent to Russian aggression, Estonia plans to mount a resistance against Russian occupation, using civilian resistance and unconventional warfare for which they are actively training.

The Estonian national security official said, "We would be happy to see the Estonian-U.S. SOF partnership furthered. Anything that relates to unconventional warfare is relevant." Estonian and U.S. SOF operated closely together in Afghanistan over the past decade, and Estonian special operations plans include a variety of support roles by U.S. SOF, scaled according to the threat level. Estonian special operations officers have described their doctrine, planning, and preparations to mount this effort, along with the types of assistance they would most like to receive from abroad. These include training, planning, and preparation initially, plus direct support in the event of escalation to armed hostilities—complete with the provision of communications links, joint terminal air controllers (JTACs), intelligence, and air support (interviews with Estonian special operations officers, June 2016).

While Estonia is preparing for a variety of scenarios, its small defense budget forces it to make some choices. Given the futility of a conventional matchup against Russia, one Estonian special operations officer said the best materiel would be man-portable air defense systems (MANPADs) and anti-tank Javelins rather than the recently purchased armored personnel carriers. MANPADS could be used to good effect by Estonians fighting a guerrilla-style war, though risks include proliferation of weapons systems that pose a threat to civil aviation if they fall into the wrong hands. Estonian SOF have closely studied the irregular tactics employed in the recent decade of fighting elsewhere to adapt them to their own needs should they become insurgents themselves. The primary injunction is for them to become a seamless part of the whole-of-government effort. An admonition to work harmoniously with civilian agencies and the population is posted in their headquarters: "No [direct action] monkeys in unconventional warfare," the poster reads, with a red line drawn through a photo of a fully kitted commando.

In July 2016, NATO announced at its Warsaw summit that a UK-led battalion would go to Estonia as part of a ramped-up four-battalion deterrence increment in the Baltics and Poland. Conversations with both Estonian and U.S. military officials indicate that the Estonians welcome the increased force presence but also want to ensure that it supports their concept and efforts. Some Estonian officials expressed concern that the U.S. forces and equipment sets planned for Poland and Germany are located too far away to be of much deterrent or practical use (interview with Estonian defense ministry officials, June 1, 2016).

In sum, this case illustrates three general principles for effectively supporting another country in measures short of conventional war:

- build a common approach;
- tap local knowledge;
- avoid redundancy or unwanted help.

Coalitions and the Evolving Role of SOF

In addition to bilateral support to a host nation confronted by aggression or coercion short of conventional war, coalitions may offer vital support. The agreed-upon measures will necessarily be the product of consensus, but that may result in a lowest-common-denominator approach. The members will sign up for those activities that are politically and practically feasible for them to perform, and in some cases they will impose limitations or caveats on their commitments. On the other hand, taking a common stand on an actual or threatened response endorsed by all members of a coalition or alliance sends a powerful message to would-be aggressors, and such signaling is an integral part of robust measures short of war.

NATO has performed this function within Europe and in out-of-area endeavors. NATO's actions as a 28-member alliance can be seen as an overarching umbrella to present a strong front against efforts by Russia, ISIL, and others to undermine individual NATO members or weaken the alliance as a whole. Nested under this umbrella are bilateral efforts, such as those undertaken by EUCOM and Special Operations Command—Europe (which is now a subunified command of

SOCOM that falls under the operational control of EUCOM). At its July 2016 summit in Warsaw, NATO moved to bolster its "reassurance" measures with more explicit forms of deterrence, including the commitment to place not just the four rotational battalions in the Baltics and Poland but also four more in Romania and Bulgaria. The more obvious it is that NATO and the United States are in a position to act on their commitments at a moment's notice, the more credible and effective their deterrent becomes. At the same time, the alliance must take care not to create or provoke the very aggression or paranoid reaction it seeks to forestall, and thus prompt war by miscalculation. Russia may, of course, claim that any such moves to strengthen deterrence, defense, and resilience are tantamount to provocations, so the diplomatic messaging is a critically important element of the response.

Through NATO SOF Headquarters, member countries' SOF have progressively developed greater interoperability over the past decade, both in the field in Afghanistan as well as in institutional development. Their singular role in responding to conflicts short of war is confined to military assistance, according to NATO SOF doctrine, which defines three categories of activities: military assistance, special reconnaissance, and direct action. The doctrine does not include support to resistance or unconventional warfare, so any such support the United States or others wish to give must be provided on a bilateral basis unless the doctrine is revised. Military assistance is typically the only activity that NATO SOF carries out in peacetime, and it roughly corresponds with the U.S. doctrinal term Security Force Assistance (SFA), which is provided to build partner capacity. However, certain activities delineated under the military assistance doctrine might be interpreted to include some support activities that the Baltic countries or other NATO members may wish the alliance to provide to supplement preparations for resistance.

Paragraph 2.2.1. of NATO SOF doctrine defines military assistance (MA) as "a broad category of measures and activities that support and influence critical friendly assets through organizing training, advising, mentoring, or the conduct of combined operations. The range of MA includes, but is not limited to, capability building of friendly security forces, engagement with local, regional, and national leadership or organizations, and civic actions supporting and influencing the local population" (NATO, 2013.)

Special MA activities include training, advising, and mentoring/partnering. Advising is a broad category that includes operational assistance, as the doctrine states: "These are activities that improve the performance of designated actors by providing active participation and expertise to achieve strategic or operational objectives." The doctrine also specifically notes that MA may be used to build a partner's preemption capabilities: "SOF do this through MA efforts to build indigenous defence and INTEL capabilities and DA on an adversary's critical operational or strategic targets" (NATO, 2013).

Finally, the doctrine calls out the particular importance of "Early Presence," that is, the proactive pre-positioning of SOF as an important factor determining their successful employment. "While a crisis is developing, SOF may be deployed to establish an early forward presence and initiate military and civilian liaison, conduct area assessments, provide an early command and control (C2) capability, or advise friendly forces. This provides the Supreme Allied Commander Europe (SACEUR) as the Strategic Commander of Allied Command Operations (ACO) and the operational-level commander with an increased understanding of a developing crisis and, if required, helps set the conditions for the initial entry of joint forces" (NATO, 2013).

Despite these elements of NATO doctrine that may permit the alliance to employ SOF in these ways, the deliberative nature of NATO decisionmaking will in all likelihood mean that its activities will remain in the realm of capacity building rather than proactive operational assistance. The early presence clause may not be invoked until a crisis has been recognized, moving NATO from its customary peacetime posture into a more robust reaction (discussions with a NSHQ official, February 1, 2016).

Individual NATO members may endorse efforts to develop a "support to resistance" concept that could eventually pave the way for a NATO doctrine that accepts resistance as a form of national defense, and allows for circumstances in which the alliance may explicitly support that form of defense. An easier path to obtaining NATO and international support, though, would be for any resistance movement to form and declare a government in exile. At that point, those wishing to support the resistance can recognize the government in exile as a legitimate government and thereby defend its assistance as support to a recognized government.

Nonetheless, support to resistance for the near term is likely to remain a bilateral effort. If the U.S. government makes a policy decision to support a resistance movement, as it has in past periods, then U.S. SOF stands ready with highly detailed, codified, and published military doctrine on unconventional warfare, including an unclassified joint doctrine published in 2015 (as well as detailed doctrine on foreign internal defense to support recognized governments). Unconventional warfare doctrine, with its emphasis on the long lead time (several years) that is typically required to assess local conditions and develop the necessary human and physical infrastructure, suggests that U.S. SOF will likely play the most active role in planning and preparing to support the resistance of any allied member countries or other partners who seek U.S. help. Even so, U.S. SOF can directly support civil defense, as in the EDL case, only when explicitly permitted to do so by the host-nation government and U.S. directives. In most cases, this will mean supporting civilian defense groups that are legally part of or reporting to the defense ministry chain of command (or, when permitted by U.S. law or policy guidance, the ministry of interior) (interview with U.S. special operations official, February 1, 2016). This is a topic of ongoing discussion, since civilian defense organizations are important thickening agents in the Baltic countries' defense schemes.

Britain as a Political Warfare Innovator

Some U.S. allies are moving of their own accord to adopt changes in policy, doctrine, and organization that will make them more capable of responding to threats short of conventional war. The United Kingdom is a particularly striking example for the array of innovations it is undertaking, partly due to military spending cuts and partly due to its assimilation of wartime lessons over the past decade. The UK army underwent a 25-percent reduction in manpower since budget cuts in 2010, but the British government has subsequently embarked on a significant recapitalization in key areas. On the theory that early, proactive, and nonlethal responses make for a cost-effective defense, Britain has invested significantly in doctrinal and organizational adaptations. It has drawn upon a long history of doctrine and organization based on political warfare, which was developed in World War II largely as a psychological warfare specialty (Garnett, 2002).

What follows is not an exhaustive treatment of the British military evolution and investment plan but a summary that highlights changes in British policy, doctrine, and organization of particular relevance to political warfare or measures short of conventional war.

First, as a matter of policy, it may be no coincidence that one of the United States' closest allies sees the current global challenges in much the same way as do the "gray zone" advocates in the United States. Britain's official National Security Risk Assessment, published in 2015, sees a growing risk of international military conflict that will draw in the UK, including through its treaty obligations. The assessment states: "Our ability to respond effectively will be made harder by the growing use of asymmetric and hybrid tactics by states, combining economic coercion, disinformation, proxies, terrorism and criminal activity, blurring the boundaries between civil disorder and conflict (HM Government, 2015)." The document also names terrorism by ISIL, al Qaeda, and its affiliates as the primary "Tier One" threat.

The UK may be ahead of the U.S. government in adopting interagency task forces as a common organizational mode for coordinating a whole-of-government response. One current example is Her Majesty's Government Counter-Daesh Task Force, which is comprised of members of all ministries and agencies, including intelligence, and housed at the Foreign and Commonwealth Office, the UK equivalent of the State Department. One UK diplomat remarked that achieving interagency coordination might be easier in a smaller government (interview, London, June 29, 2016).

Second, at the level of military doctrine, Chief of General Staff General Sir Nicholas Carter has internalized his extensive experience in Afghanistan to embrace the need for a wider view of maneuver to encompass information and other dimensions, as well as the need for adaptability among smaller forces to wage the full spectrum of conflict.[8] British forces will also, furthermore, most often act as part of a larger coalition, requiring adaptability to achieve interoperability. The UK army has developed doctrine calling for "integrated action" that combines maneuver, fires, information activities, and security capacity building as needed for complex conflicts, which it sees as the dominant form of conflict today. An internal UK army note was published last year, outlining the new doctrine to be published in its forthcoming capstone publication.

Information activities assume a particularly large and expansive dimension in British thinking about warfare today. An officer involved in developing the new army doctrine summarizes the thinking behind it: "We have to understand and take account of all actors. The population is critical to the outcome. There is no definitive victory and no clear end point such as a [formal] surrender. World opinion will determine if we won" (interview with UK army officer, London, June 27, 2016). In addition to Britain's own experiences, he offered the example of Israel's struggle with world opinion over its recent conflict in Gaza. "Israel lost the battle of perception and therefore lost in Gaza," he said.

The British doctrine also calls for greater synergies with nonmilitary forces. "We need to blur the edges of capabilities," the officer said. "If a state wants to use its tools, it makes sense to do it cooperatively and with as few gaps as possible." He cited the UK military's recent deployment under a civilian "two star" to Sierra Leone as part of the international assistance to African countries suffering from the Ebola epidemic as an example of the types of adaptation in command structures that he considers likely in the future.

British history is replete with examples of limited military operations that involved little to no conventional combat, he noted, recalling his own father's deployment to Aden, Yemen. "We used to call them Military Operations Other Than War (MOOTW), and we applied

---

8 For complete remarks, see transcripts of the Chief of General Staff speeches on adaptability and integrated action, delivered at the annual UK land warfare conferences in 2016 and 2015, are posted at the Royal United Services Institute website. As of Janauary 18, 2017: https://rusi.org/landwarfare

it in the Empire and post-colonial policing, and punitive actions, as opposed to [the] Crimea expedition to conduct war with a peer against another peer. These lesser cases are actually most of what we have done, but after the Cold War we chose to maintain forces and doctrine to deal with those occasional existential threats. Now we see that we have a revisionist power, but the most demanding task is not necessarily fighting another state but the irregulars mixed in with the civilians in the Ukraine." He cited another example from the end of World War II, when 20,000 Greek Communists began marching toward Athens to seize it before the government in exile returned. The UK sent three infantry battalions to seize the high ground and key government installations to preempt the Communist takeover. It built up the Greek army and police, set up soup kitchens, sent a Greek unit to Crete to prevent it from committing atrocities in revenge, and flew British trade unionists in to verify and publicize that the 18-month operation did not constitute a repression of Greek citizens.

Third, the UK army has also embarked on a novel effort to retool some of its organizational structure and personnel for nontraditional missions. The locus of this activity is the 77th Brigade, a joint brigade that General Carter created following his last tour in Afghanistan.[9] The transformational efforts of the UK army aim to remedy deficits it has discovered in 15 years of confronting irregular threats, as it fought in coalitions alongside the U.S. For example, Carter led British troops in Shi'a-dominated southern Iraq, where they encountered Iranian-backed militias and complex political machinations. One of his senior officers recalled how ill-equipped the coalition was to address this type of warfare. He said, "In Basra, we knew [we] were fighting the wrong enemy with the wrong tools—.556 ammo and tanks. The Badr Brigade [the dominant Shi'a militia with deep Iranian ties] was a veterans organization with a foreign policy, with political, economic and security lines of operations. We knew this but did not know how to address it. CPA [the Coalition Provisional Authority] wrote policy

[9] The official website of the 77th Brigade is http://www.army.mod.uk/structure/39492.aspx.

papers. . . . In Afghanistan we encountered the same thing" (interview with UK army officer, June 27, 2016).

This officer views the current transformation ordered by Carter as critical to confronting the current threats, and he emphasizes how profound this change is for an organization that has been oriented largely to conventional warfare since the cold war. "The infantryman is conditioned to simplicity, to remove doubt, to act swiftly to kill in combat. This is entirely different from the gray areas in which we find ourselves," he noted. In his view, the work on unrestricted warfare published by two Chinese colonels in 1999 constitutes the most insightful treatment of the modern mode of warfare, which conceives of the world as the battlespace in which close and deep fires are aimed at domestic and distant audiences, with time being a critical element.

To address the needed gaps, Carter created the 77th Brigade, which is a novel formation that is deployed in support of other military formations to assist them in conducting nontraditional warfare. This model for transforming military operations is intended to preserve conventional fighting competence but to add tailored capabilities as needed. The brigade includes a wide range of competencies that are currently located in U.S. SOF as well as some unique skills imported from the private sector and civil society. The brigade delivers both nonlethal military specialties and nonmilitary specialties, which are task organized into mission groups and smaller eight-person hybrid warfare teams. The nonlethal military specialties include psychological operations/media operations, civil affairs, and security capacity building at the ministerial or institutional (as opposed to tactical) level. The nonmilitary capabilities are specialists in political operations (i.e., understanding the politics of other countries), business, finance, informal economy, organized crime, nongovernmental and international voluntary organizations, and proxy forces.

When Brigadier Alastair Aitken assumed command of the 77th Brigade two years ago, he first set about defining the personnel requirements for the new unit. Working with British Special Forces, he developed selection criteria based on lateral thinking, creativity, emotional intelligence, and all types of communications skills, for both active-duty and reserve recruitment. He built upon a British reserve model whereby senior business executives serve as top advisers to the military in areas of engineering and logistics as a special reserve. He obtained exemptions to allow him to recruit reservists with unique skills and recognition in technology, creative arts, and the above specialties—and to transform their terms of service. They may work remotely or part time, and they may provide reachback services to deployed troops. Similarly, he has arranged to place troops in training fellowships at companies or other organizations where they can gain discrete skills and build social networks to aid the brigade. "We want people who think differently, who can tear ideas apart, an idea factory," an officer at the 77th Brigade explained. The Hybrid Warfare Center at Dennison Barracks is designed as a brainstorming center for a core of the 500-strong joint brigade, 40 percent of whose troops are reservists. The nascent effort is now supporting UK and coalition forces in Iraq and Africa. While the effectiveness of these innovations will require some time to determine, and very likely some additional refinement, the experimentation and organizational flexibility are useful models for emulation.

Recommendations for Military and SOF Support to an Integrated Approach

The employment of all elements of national power to counter threats short of war is, in the first instance, a description of effective statecraft. For diplomacy to succeed, it must be backed by sufficient leverage, which is the essence of political power, as Hans Morganthau defined it. Political power emanates from the possession of sufficient military, economic, and reputational clout to achieve ends *without* resorting to force. Such power can be exercised through threats or incentives, and most likely a combination of the two. In this view, military power is a supporting element in the effort to achieve objectives without going to war. Many displays of military might and intent fall into the realm of signaling through presence, posture, exercises, and other shows of force that will involve either primarily conventional forces or a combination of conventional and SOF.

Special operations forces are particularly suited to a multitude of roles in the political warfare realm, due to their diverse capabilities and to their ability to operate at low levels of visibility in semipermissive environments or behind enemy lines. The military, and primarily SOF, can play three overarching roles in responding to the challenges short of war. First, in sensitive, semipermissive, or nonpermissive environments, SOF are trained to help gather and synthesize information about ambiguous and possibly threatening developments that might signal subversion, coercion, or other attempts to destabilize critical regions, allies, or friends. Given the clandestine and subtle nature of these threats, a great premium for any effective response must be placed on early efforts to detect and understand what may be happening. Second, civil affairs and MISO can play a significant role in enhancing the resilience of targeted societies. These two SOF capabilities are already in use in many embassies to support diplomacy with host nations by addressing socioeconomic conditions that generate conflict and meeting informational needs. In some cases, specific authorities or permissions may be needed to address conflicts that do not involve terrorism. Third, SOF can help prepare for, and support, national resistance to subversion and coercion; SOF aid in these missions by working with host-nation security forces as well as informal forces in countries where governance is weak, nonexistent, or malign. Some European states' defense strategies include reliance on national and civil resistance to counter subversion or invasion. Ideally the SOF capabilities should be available to support those resistance efforts where the host nation desires such support. In the case of the current fight against ISIL, the United States and coalition partners are working with an array of informal forces in Syria and Iraq, as they have also done in Afghanistan, Somalia, Libya, and elsewhere.

This final section draws upon interviews and other research to synthesize a number of implications and recommendations for the U.S. military and SOF in particular that may be useful in improving strategy, planning, and execution across the government to successfully mitigate, manage, or defeat aggression short of conventional war. These military capabilities are likely to be employed as part of a whole-of-government strategy, given the nature of political warfare as characterized in this study. The above sections indicate that the leading role may most often be played by civilian entities, other countries, and possibly entire coalitions. Thus, the U.S. military and SOF roles in particular will necessarily need to nest closely with those other actors. It follows that many of the implications and recommendations emphasize a greater degree of collaboration with other governmental actors to understand the environment and form a plan of action. Other implications emphasize the military or SOF capabilities needed to address a diverse and increasing array of challenges short of war. At the same time, some of the challenges may not require any military role whatsoever.

While many military activities in support of a largely civilian response to nonconventional threats are not controversial, some areas of SOF activity may give rise to controversy, as they may risk disrupting diplomatic relations or triggering an escalation of crises. And in some areas the U.S. government has only rarely authorized SOF activity, such as in supporting resistance activities inside other countries. Policymakers will necessarily weigh risks and benefits in arriving at the most effective responses. In addition, developing and applying measures of effectiveness for the range of special operations activities will give policymakers greater confidence of the results they can expect to achieve.

Recommendation 1: To improve whole-of-government synergy, U.S. military commands, including deployed headquarters, should as a matter of course include civilian departmental representatives in order to understand, coordinate with, and support U.S. State Department and other civilian program execution. The extension of civilian reach—as a formal task—should be directed from national decision makers and written into the formal mission statements and plans of military organization as a requirement rather than an ad hoc or exceptional occurrence. Most efforts will benefit from a combined civilian-military effort, and organizations should routinely be formed with a diverse set of personnel. Such collaboration may need to be written into the instructions for military task forces and commands. In cases where U.S. missions or consulates are inadequate, the execution of civilian missions may require the use of military platforms and organizations. The U.S. military has a large number of platforms in the form of task forces, military commands, and bases around the world. Thus, it may be that these platforms will serve as the physical locus for coordination of what could be nonmilitary or civilian-led efforts.

Over the past decade advances have been made. Over 80 foreign policy advisers and a similar number of USAID representatives are now assigned routinely to major military commands, but operational headquarters and task forces do not always include such personnel. A State Department official stated that, in her experience, the special operations community has been in the forefront in recent years of forming interagency task forces and welcoming civilian partners in efforts being conducted in Africa, South Asia, the Middle East, and elsewhere. "It's in SOF's DNA to take a civilian-military approach," she said. "This provides a way to reach a common diagnosis and a military-diplomatic approach to addressing it" (interview, April 12, 2016). But other entities have not always been so welcoming. Despite a plan approved for State and DoD to tackle a failed state's problems together, for example, a regional task force was unwilling to host a State Department team.

Recommendation 2: The Defense Department, and special operations forces in particular, should incentivize and improve selection and training for military advisers serving at State Department headquarters, U.S. embassies, and other diplomatic posts in order to increase their effectiveness. The U.S. military has for years assigned military advisers to the State Department, and more recently special operations personnel have been assigned as well, to Special Operations Support Teams (SOSTs) in Washington and to SOF Liaison Elements (SOFLEs) in embassies. The quality of the military adviser and his or her preparation can be determining factors in whether a given effort achieves the needed level of integration among the military and civilian entities. An astute adviser will understand the culture and process of the State Department and be proactive in helping his military colleagues navigate it successfully. He or she will also lean forward to ensure that State is informed of and invited to critical meetings convened by the military.

A particular problem concerns the placement of special operations personnel in embassies, where the normal tour for members of the country team is two years or longer. The special operations community, small and operationally oriented, is reluctant to assign advisers for longer than a year. The career path for upwardly mobile officers creates distinct disincentives for a two-year tour, and the result is that few star officers wish to serve in those positions. At a minimum, the SOF leadership needs to understand and take account of the impact this may have on the success of the SOF liaison within the country team, and seek ways to mitigate the negative impact and increase the liaison's ability to offer value. USSOCOM has been willing to pay for the cost of housing a SOFLE at missions, but the issue goes beyond the financial burden on State's much smaller budget. One interviewee suggested that liaisons should all serve a minimum one-year tour, and another suggested that the military services' personnel models should incentivize this vital function. High-quality senior officers nearing the end of their careers who will make a final three-year commitment may be the ideal candidates for these positions.

The final issue regarding military advisers involves the chain of command: While a SOFLE continues to report to his or her military command, he or she must treat the embassy chief of mission as his or her boss as well. This deference to the chief of mission requires significant diplomatic skill and, in emergency circumstances, the intercession of the higher command to ensure that any disagreements are amicably resolved. The SOFLE must establish a constructive relationship with the senior defense official at the mission. If the liaison is not seen as a trusted member of the team, his or her utility will be limited.

Recommendation 3: The Defense Department, and special operations forces, should offer military planners to the State Department as the latter builds its own cadre of planners and increases coordination among regional and functional bureaus, to enable State to play a lead role in responses to political warfare. While the State Department enhances its own planning capacity, the U.S. military could formally second its trained planners to each State Department regional bureau and the policy planning bureau to provide a dedicated source of manpower to develop political-military plans that would complement the military theater campaign plans and, ideally, lay the foundation for those plans. These documents could serve as the forcing function for implementing a DIME approach to threats short of conventional war. The documents could fill the gap between the normal diplomatic goals articulated in the Integrated Country Strategy and the military suite of plans that flow from the National Military Strategy, the Guidance for the Employment of the Force, and other military guidance. With the development of a set of plans on a routine basis, the expression "whole-of-government" will cease to be an aspirational phrase and will acquire a concrete form for measuring progress, effects, and areas for adjustment.

Recommendation 4: Military commanders should develop and maintain collaborative relationships with their civilian counterparts through regular visits and frequent communications to develop common understanding of and approaches to political-military conflicts. The military leadership should not rely solely on liaison and advisory personnel but, in addition, seek regular visits and make regular calls to key personnel at the State Department and embassies. One diplomat suggested that quarterly visits would be ideal to develop and maintain personal relationships with key civilian officials; and weekly or biweekly calls, to maintain communications on significant issues. This effort to resolve differences and develop a combined approach will provide direction to their subordinates and allow them to focus on detailed planning and execution. Without agreement at the top of the respective State and DoD structures, no coordinated U.S. government approach can be achieved. This agreement should logically precede engagement with the host country government to develop substantive plans and agreements.

Recommendation 5: DoD should routinely seek to incorporate State Department knowledge and the current insights of the U.S. country team into military plans in order to develop effective responses to political-military threats. The country team constitutes the most valuable source of current authoritative knowledge that exists within the U.S. government about a particular country. The U.S. country team members are constantly in touch with a wide array of actors in its assigned country, and one of its primary missions is to understand and report on significant political, economic, cultural, and security developments. The U.S. military, including SOF, can exploit this source of knowledge more actively in conducting assessments, formulating military plans, and collaborating on government-wide plans. The theater campaign plan, the embassy's integrated country strategy, and the regional bureaus' plans should not be formulated in isolation from one another. U.S. military plans are often framed in extremely broad language and do not take into account critical information and holistic assessments of a country's political dynamics. The development of threat scenarios and likely courses of action will be greatly improved by more extensive input.

Recommendation 6: The special operations community should make it a high priority to improve and implement fully resourced, innovative, and collaborative information operations. Extensive research, training and collaboration are required to improve the foundations and methods of this key discipline, as described in detail in Chapter Eight. Military Information Support Teams (MISTs) and Civilian Military Support Elements (CMSEs) are in high demand from many chiefs of mission, but there remains substantial unease in the diplomatic corps about perceptions of the "militarization of foreign policy" and the potential for unintended disruption of delicate relations during information operations in particular. U.S. diplomats are naturally wary of being accused of interference in another country's internal affairs, as this can compromise their ability to function, sometimes severely. In past years, unfortunate diplomatic incidents have occurred when U.S. military personnel entered countries without embassy-granted "country clearance" or while gathering information in sensitive areas. A prolonged diplomatic crisis ensued when Pakistan detained a CIA contractor after he shot individuals who had accosted him and stolen his surveillance equipment. These risks always exist, and U.S. intelligence and other activity should not cease simply because they may run afoul in some way. Nonetheless, steps could be taken to address the unease and achieve greater integration.

For example, an experiment could be conducted to form blended civilian-military teams to carry out information operations. This would represent a significant departure from U.S. SOF's normal procedures, but it mirrors similar steps taken in recent years to develop Cultural Support Teams and to partner with Spirit of America volunteers. As the UK's 77th Brigade is doing, perhaps an experiment to field blended Community Service Officer (CSO), USAID, and CMSE teams would be beneficial. MISTs already routinely work in embassies' public diplomacy sections, but perhaps a further step could be to form standing teams of State public diplomacy and MIST personnel.

If these models prove successful, the demand for MIST and Civil Affairs personnel could grow in coming years, necessitating an expansion of their ranks. Currently, these small and heavily deployed units are just returning to a normal operational tempo. Any surge requirements would place them under strain again. They are perhaps the most valuable tactical contribution that SOF can make to whole-of-government efforts to counter political warfare. CMSEs recently deployed in a country conducted critical assessments of Russian national youth populations, revealing a very important distinction between their attitudes' toward the West and their parents more hardline attitudes. Such insights can provide the basis for a nuanced and effective outreach program.

Finally, significant additional resources, education, and training should be provided to create a new generation of psychological operations (PSYOP) talent. MIST or PSYOP capabilities used to be firmly grounded in a tactical suite of radio broadcast and print leaflet operations. Today, those capabilities may still be necessary, but in addition, a major investment is needed to ensure that MIST becomes as expert in using social media and other cutting-edge information technologies as any entity in the U.S. government. These capability requirements span analytic tools, collection mechanisms, and delivery models. The entire U.S. military places a constant demand on the intelligence community for a wide variety of assets, including all-source analysts. MISO relies heavily on intelligence to conduct precision target audience analysis. However, satisfying the MISO requirements are far down the priority list when they should be at or near the top, if the information arena is indeed the critical battlefield. Intelligence support to MISO may be the most pressing requirement for developing a world-class military information capability that can lend support to civilian and whole-of-government organizations.

Recommendation 7: Military commanders and the State Department should identify critical information requirements for political warfare threats, and the intelligence community should increase collection and analysis capabilities that are dedicated to detecting incipient subversion, coercion, and other emerging threats short of conventional warfare. The policymaking community has awakened to the need to develop a range of collection priorities of this type, as evidenced by the recent request for intelligence reporting on Russian economic aid to European political leaders, parties, and other entities (Foster, 2016). Tracing financial flows is a promising but difficult method to understand how and where countries are seeking influence for political purposes. Additional policy and legal issues must be navigated to determine legitimate targets of inquiry and permissible procedures for collecting information. However, the mapping of open source information about the influence networks of adversaries and potential adversaries is necessary to understand how political power is being amassed and wielded. Determining when those networks are being used for subversive, coercive, or aggressive purposes is necessary if sufficient warning is to be supplied to policymakers. Attributing cyber attacks is another area requiring technical resources and enhanced policy focus. Finally, unmasking adversary measures without compromising intelligence sources and methods poses an additional related challenge for policymakers when threats warrant countermeasures.

Recommendation 8: DoD and State should support deployment of special operations forces in priority areas deemed vulnerable to political warfare threats as an early and persistent presence to provide assessments and develop timely and viable options for countering measures short of conventional war. Interviews with SOF have stressed this requirement above all others: Their ability to assess conditions and formulate and validate options for policymakers requires a presence on the ground in emerging trouble spots. The initial step required by doctrine is an assessment of the conflict and of the potential partners' objectives, needs, and capabilities. While intelligence collection and analysis can be done remotely to a large degree, both the operational assessment and the formulation of concepts of operation require SOF to conduct physical inspections and face-to-face meetings. Otherwise, the U.S. government cannot expect the special operations community to provide sound and detailed options in a timely fashion. While military commands and DoD initiate the proposals for such deployments in nonwar zones, the U.S. chief of mission authorities grant country clearance for U.S. personnel to enter countries. Thus, both State and DoD support for this use of special operations forces is imperative.

A proactive stance may provide policymakers with more options, and risk mitigation strategies can alleviate downsides. The interviewees cited numerous cases in Eurasia, Syria, and Iraq where the options for countering Russia, ISIL, and other potential adversaries could not be developed in a timely manner because commanders were not willing to authorize site visits or contacts with exiles or expatriates. In the case of unconventional warfare or support to resistance, the interviewees stated that a minimum of three to five years is needed to conduct an assessment and, if warranted and approved, to build the needed infrastructure to support resistance. Policymakers must recognize that these missions are likely to go awry if ordered up without sufficient lead time. In addition, assessments should evaluate the likelihood that activities will be exposed in the ubiquitous global media environment. Proposals should include measures to mitigate exposure of any authorized sensitive, clandestine or covert activities. In some cases the effects of exposure may outweigh the benefits of such operations.

References—Chapter 8

Boot, Max, and Michael Doran, "Political Warfare," *Policy Innovation Memorandum No. 33*, Council on Foreign Relations, June 7, 2013.

Blackwill, Robert D., and Jennifer M. Harris, *War by Other Means: Geoconomics and Statecraft*, Cambridge, Mass: The Belknap Press of Harvard University Press, 2016.

Brands, Hal, *What Good Is Grand Strategy? Power and Purpose in American Statecraft from Harry S. Truman to George W. Bush*, Ithaca, N.Y.: Cornell University Press, 2014.

Bremmer, Ian, *Superpower: Three Choices for America's Role in the World*, New York: Portfolio/Penguin, 2015.

Broadcasting Board of Governors, *About Us*, Washington, D.C., n.d. As of March 8, 2018:
https://www.bbg.gov/wp-content/media/2011/10/BBG_FACTSHEET_v20_single-pgs-10-7-15.pdf

Connable, Ben, Jason H. Campbell, and Dan Madden, *Stretching and Exploiting Thresholds for High-Order War: How Russia, China, and Iran Are Eroding American Influence Using Time-Tested Measures Short of War*, Santa Monica, Calif.: RAND Corporation, 2016.

Cordesman, Anthony, "Russia in Syria: Hybrid Political Warfare," *CSIS*, September 23, 2015. As of November 30, 2015:
http://csis.org/publication/russia-syria-hybrid-political-warfare

Foster, Peter, "Russia Accused of Clandestine Funding of European Parties as US Conducts Major Review of Vladimir Putin's Strategy," *The Telegraph*, January 16, 2016. As of July 14, 2016:
http://www.telegraph.co.uk/news/worldnews/europe/russia/12103602/America-to-investigate-Russian-meddling-in-EU.html

Garnett, David, *The Secret History of PWE: The Political Warfare Executive, 1939–1945*, Boston: Little, Brown, 2002.

Gompert, David C., and Hans Binnendijk, *The Power to Coerce: Countering Adversaries Without Going to War*, Santa Monica, Calif.: RAND Corporation, 2016.

HM Government, "Summary of the National Security Risk Assessment 2015," Annex A to the National Security Strategy and Strategic Defence and Security Review 2015: A Secure and Prosperous United Kingdom, HM Government, November 2015. As of January 18, 2017:
www.gov.uk/government/publications

Inside.gov, United States Budget Estimate, at InsideGov, 2016. As of July 9, 2016:
http://federal-budget.insidegov.com/l/119/2016-Estimate

Kralev, Nicholas, "Congressional Staff AttitudesToward the Foreign Service and the Department of State," prepared by Nicholas Kraleve for the American Foreign Service Association, June 2013, p. 15. As of January 18, 2017:
http://www.afsa.org/sites/default/files/Portals/0/2013_Congressional_Attitudes.pdf

National Security Strategy and Strategic Defence and Security Review 2015: A Secure and Prosperous United Kingdom, HM Government, November 2015. As of January 18, 2017:
https://www.gov.uk/government/uploads/system/uploads/attachment_data/file/478936/52309_Cm_9161_NSS_SD_Review_PRINT_only.pdf

Kissinger, Henry, *World Order*, New York: Penguin Books, 2014.

Mearsheimer, John L., *The Tragedy of Great Power Politics*. New York: W. W. Norton, 2014.

Morganthau, Hans J., *Politics Among Nations: The Struggle for Power and Peace.* New York: McGraw Hill. Seventh Edition, 2006.

NATO. AJP 3.5 Allied Joint Doctrine for Special Operations, Edition A, Version 1, December 2013. North Atlantic Treaty Organization.

Nye, Joseph S., *Soft Power: The Means to Success in World Politics*. New York: 2004.

Parameswaran, Prashanth, "Beware China's Political Warfare Campaign Against US, Allies: Experts," *The Diplomat*, October 23, 2015. As of November 30, 2015: http://thediplomat.com/2015/10/beware-chinas-political-warfare-campaign-against-us-allies-experts/

Posen, Barry R., *Restraint: A New Foundation for U.S. Grand Strategy*. Ithaca, N.Y.: Cornell University Press, 2014.

Ries, Charles P., *Improving Decisionmaking in a Turbulent World*. Santa Monica, Calif.: RAND Corporation, PE-192-RC, 2016.

CHAPTER NINE

Conclusion

Emile Simpson ends his book *War from the Ground Up* by cautioning against the blurring of lines between war and politics: "War is currently not compartmentalized as it should be. By increasingly merging it with regular political activity, and investing in operational ideas a policy-like quality, we are confronted with policy as an extension of war; that used to be the other way around." (Simpson, 2012, pp. 243–44.) He is referring here to the Clausewitzian formulation that war is an extension of politics (policy, as Simpson translates it).

Despite Simpson's plea, it is likely today that the continuum is blurrier than ever between peace and war, between political contests that are unarmed (but may involve an array of other coercive measures) and those that are armed. It may not be our choice to blur, but the reality in which we live. The choices of what to call things are still ours to make, however. The naming of types of conflict should clarify rather than confuse, and also represent the reality accurately.

It would not do to turn our back on "armed politics" as a phenomenon, since it is being so widely practiced by state and nonstate actors on all sides. As Simpson himself argues, "The use of armed force within a traditional conception of war on the one hand, and outside war, as 'armed politics' in a direct extension of policy, on the other, is an important distinction in the abstract. . . . The failure to distinguish [between these two] carries with it a risk of liberal powers potentially getting the worst of both parts: engagement in permanent conflicts in whose operational prosecution they are unskilled" (Simpson, p. 11).

The United States and its allies do not have the luxury of choice in modes of warfare, as enemies will choose to fight in the ways most advantageous to them. Most often, they will, if they can, use a blend that Frank Hoffman dubbed hybrid warfare (Hoffman, 2007). This trend was first noted by Martin van Creveld in his work *The Transformation of War* (1991). On grounds that the attributes of conventional warfare are well known, well studied, and codified, this monograph has focused on the attributes of irregular warfare. The attributes identified through our case studies and listed in Chapter Seven serve as a descriptive starting point, and subsequent chapters identify the key elements of an effective response and the key changes the U.S. government needs to make to implement that response.

This study further argues that developing nonlethal measures to respond to these challenges, and more effectively orchestrating those measures under the State Department's oversight, is the most prudent and cost-effective strategy for safeguarding the nation's interests while avoiding unnecessary and often unproductive military interventions.

References—Chapter 9

Hoffman, Frank G., *Conflict in the 21st Century: The Rise of Hybrid Wars*, Arlington Va.: Potomac Institute for Policy Studies, 2007.

Simpson, Emile, *War from the Ground Up: Twenty-First-Century Combat as Politics,* London: Hurst and Company, 2012.

Van Creveld, Martin, *The Transformation of War: The Most Radical Reinterpretation of Armed Conflict Since Clausewitz*, New York: Free Press, 1991.

APPENDIX A

Defining Political Warfare

Table A.1
Political Warfare's Multiple Definitions

Author	Source	Definition
George Kennan	"On the Inauguration of Organized Political Warfare," 1948, paragraph 1	Political warfare is the logical application of Clausewitz's doctrine in time of peace. In broadest definition, political warfare is the employment of all the means at a nation's command, short of war, to achieve its national objectives. Such operations are both overt and covert. They range from such overt actions as political alliances, economic measures (as ERP—the Marshall Plan), and "white" propaganda to such covert operations as clandestine support of "friendly" foreign elements, "black" psychological warfare and even encouragement of underground resistance in hostile states.
Paul Smith	*On Political War,* 1989, p. 3	Political war is the use of political means to compel an opponent to do one's will, *political* being understood to describe purposeful intercourse between peoples and governments affecting national survival. Political war may be combined with violence, economic pressure, subversion, and diplomacy, but its chief aspect is the use of words, images, and ideas, commonly known, according to context, as propaganda and psychological warfare.
Angelo Codevilla	"Political Warfare," in *Political Warfare and Psychological Operations: Rethinking the US Approach*, 1989, p. 77	Political warfare is the marshaling of human support, or opposition, in order to achieve victory in a war or in unbloody conflicts as serious as war.

Table A.1—Continued

Author	Source	Definition
Carnes Lord and Frank Barnett	"Introduction," in *Political Warfare and Psychological Operations: Rethinking the US Approach*, 1989, p. xi	Political warfare is a term that is less well established in usage and doctrine, but one that seems useful for describing a spectrum of overt and covert activities designed to support national political-military objectives.
Carnes Lord	"The Psychological Dimension in National Strategy," in *Political Warfare and Psychological Operations: Rethinking the US Approach*, 1989, p. 18	Political warfare is a general category of activities that includes political action, coercive diplomacy, and covert political warfare.
United States Army Special Operations Command 2015	"SOF Support to Political Warfare White Paper," March 10, 2015, p. 2	Political Warfare encompasses a spectrum of activities associated with diplomatic and economic engagement, Security Sector Assistance (SSA), novel forms of Unconventional Warfare (UW), and Information and Influence Activities (IIA).

Lightning Source UK Ltd.
Milton Keynes UK
UKHW020720230721
387619UK00005B/84

9 780833 097071